Environmental Science and Engineering
(For B.E./B.Tech. Students)

Dr. C. Paul Raj

Assistant Professor

University College of Engineering Villupuram

(Constituent College of Anna University of Technology)

Chennai.

Published by

BONFRING®
Intellectual Integrity

ISBN 978-93-84743-64-2

Author

Dr. C. Paul Raj

Bonfring

309, 2nd Floor, 5th Street Extension, Gandhipuram,

Coimbatore-641 012.

Tamilnadu, India.

E-mail: info@bonfring.org

Website: www.bonfring.org

Phone: 0422 4213231

Preface

It is the sincere effort to introduce the concept of "Environmental Science and Engineering" to Engineering pupil. The book strictly covers all the aspect of syllabus prescribed by Anna University of Technology. Relevant and important case studies have been presented. The pictorial representation of concepts with lucid presentation adds new value to the book. The book contains five units each with sub-titles. Each unit corresponds to the syllabus by the university. In relevant places historic developments is included to make the theory interesting. Where ever possible important points are bulletin numbered to minimize the time of reading lengthy theory.

The best way to use the book is to see the syllabus word and then go to the page. The same word appears and the page number for the text matter can be identified from there. In variably, the content in one unit is discussed in the same unit as for as possible arranged sequentially. It is to be noted that no part of the syllabus is omitted from the discussion. Keeping in few of the students environment questions are provided at the end of the each chapter.

A comprehensive account of short question answer followed by a glossary has been included in this book.

As per the exception of University curriculum recent developments and incidents in environmental issues has also been included in relevant places of the text. Important Indian cases like earthquake (Gujarat), Bhopal Tragedy, Tsunami, etc are included.

It is sincerely felt that the mission to make the present engineering generation ecosensitive and enjoy Environmentology. The author welcomes constructive criticism of any kind and also the queries for any assistance.

Dr. C. Paul Raj

UNIT I

ECOSYSTEM AND BIODIVERSITY

Definition, scope and importance of environment – need for public awareness - concept of an ecosystem – structure and function of an ecosystem – producers, consumers and decomposers – energy flow in the ecosystem – ecological succession – food chains, food webs and ecological pyramids – Introduction, types, characteristic features, structure and function of the (a) forest ecosystem (b) grassland ecosystem (c) desert ecosystem (d) aquatic ecosystems (ponds, streams, lakes, rivers, oceans, estuaries) – Introduction to biodiversity definition: genetic, species and ecosystem diversity – bio geographical classification of India – value of biodiversity: consumptive use, productive use, social, ethical, aesthetic and option values – Biodiversity at global, national and local levels – India as a mega-diversity nation – hot-spots of biodiversity – threats to biodiversity: habitat loss, poaching of wildlife, man-wildlife conflicts – endangered and endemic species of India – conservation of biodiversity: In-situ and ex-situ conservation of biodiversity.

Definition of Environment

Environment is the sum of all social, economical, biological, physical and chemical factors. It constitutes the surrounding of man and other living and non-living forms. The term environment was introduced by Jacob van Uerkul. The word 'environment', is derived from the French word 'Environ' means encircle or surround. Environment is made up of living entities, non-living part of the environment and micro components Light, temperature humidity, etc. Natural environments are Forest, lake, sea, etc. Artificial Environment are ship, aircraft, etc.

Scope of Environmental Education

A comprehensive knowledge in environmental educations helps in environment management. Few are listed below.

- Conservation of Biological Diversity.
- Conservation of Land and Forest.
- Disaster Management.
- Conservation of Energy Resources.
- Sustainable Development.
- Environmental law and its importance in environment sustainability.

Role of Youth

It is necessary to enrich the minds of youth regarding environment to make a better tomorrow. Younger generation must concern in environmental sustainability. They have to be properly educated in sustainable use of resources.

Role of Professionals

Professional engineers and scientists must involve themselves in environmentally sustainable development.

(i)   Develop technologies and processes, which are environment friendly.

(ii) Detect the environmental degradation and take corrective action.

(iii) Implement 'zero waste', 'waste reuse' strategies in all levels of life.

(iv) Use resources sustainable and help to maintain an ecological balance.

(v) Coordinate and cooperate with different environmental agencies and contribute to environmentally sustainable development.

(vi) Have a motto of "think globally and act locally".

Limitations of Environmental Education

Population Explosion: Better environment leads to better life and increased life time. Longer living results in more utilization of resources. It affects the needs of potential future generation.

Today almost every organ of one human can be transplanted and hence the effective life time of human body is further increased.

Just to Fit: Education leads to knowledge. Knowledgeable man knows the nuke and corners of environmental requirements. Cunning man adopt methods, which just satisfy or appears to satisfy the requirements.

Capital: Practice of environmental sustainability in advanced level is capital intensive. Many companies hesitate to implement their knowledge in environmental requirements.

Hurdle to Development: Man wants to achieve maximum in minimum time and using innovative q methods.

The environmental concepts further include complexity in management principles and concepts. Industrial development with sustainable concept is a slow process.

Imbalance: Developed countries are in general non-cooperative though they may make themselves as if involved in environmental concern.

Indeed, many developed countries exploit developing and under developed countries to manufacture chemicals, which pollute the environment.

Importance of Environmental Education

It is an important tool to educate the people for preserving quality environment. Educational curriculum, media like television, radio, newspapers, conducting workshops, etc. improves importance of environment.

- The environment that nurtured us has become our heritage. Preserving the same is our individual ethical duty.
- Develop consciousness of my earth is my backyard and its health and wealth is an important personal concern.
- Gain the knowledge of what is right to environment and what is not. The knowledge of different type of environment and the effects of different environmental hazards.
- The knowledge provides better choice of environmental life and which earns better future especially in the last half of life from environmental diseases.
- Different type of pollutants and it effect to environment can be learned.
- Disasters and its management is another area of important environment education.

Need For Public Awareness (Environmental Education)

India is committed to be an environmental friendly state by its Constitution and also by signing United Nations Earth Submit held in Rio de Janeiro, Capital of Brazil on 3rd June, 1992.

Government of India has accepted the recommendations of Tiwari Committee Report (1980). It emphasized the need for environmental education.

The suggestion includes both formal and non-formal education. The formal includes four distinct but inter related components for school students up to higher secondary education.

Further Supreme Court of India has directed the State to provide environmental education. and through curriculum (M.C. Mehta vs Union of India, 1988).

Steps Taken by Government of India

- Introduced formal and non-formal programs on environment and its importance.
- Enviroment Science, Environment Engineering, etc are initiated as specialized area of study.
- Department of Environment (1982), Environmental Information Systems (ENVIS), Center for Environmental Education (CEE) are some of government organs involved in environment management.
- It proclaimed many environment laws.

- There are more than 200 non-government organizations.The environmental protection Society has different institutes to propagate environmental education.

Concept of an Ecosystem

The inter and intra relationship among both biotic and abiotic components is studied under ecosystem ecology. The term "ecosystem" was coined by A.G. Tansley in 1935. He defines as "the system resulting from the integration of all the living and non-living factors in the environment". Ecosystem is the basic functional unit of ecology.

Natural and Artificial (Man-Made) Ecosystem

The ecosystem existing due to natural evolution of earth is a natural ecosystem. Eg. forest, grassland, desert, rivers, sea, ponds, lakes, etc.

Artificial ecosystem is the one which is planned by man in recent times. Eg. aircraft, submarine, ship etc,

Components of an Ecosystem: Components are intangible and intangible resources constituting ecosystem.

Biotic Component: All living plants and animals constitute biotic components

Abiotic Component: All non-living tangible things form abiotic component. Eg. Sand, water, etc.

Energy Component: It is the interconnecting component between biotic and abiotic components. Sun is the ultimate source of energy. Biotic components consume energy for life processes.

Characteristics of an Ecosystem

- Major structural and functional unit of ecology.
- Structure is the species diversity is inherent in the ecosystem.
- Functional component includes energy flow and material cycling.
- Matter and energy flows from less complex to complex states in the ecosystem.
- Ecosystem interact with environment for energy balance and stability.

Structure of an Ecosystem

The term 'structure' refers to the various components and their relations in the ecosystem.

- Composition of biological (biotic) community, which includes plants, animals and microbes and their relative distribution.

- Interrelation between biotic community with non-living (abiotic) materials.
- Microclimate conditions and their variations like temperature, light, humidity, topology, etc.

Function of an Ecosystem

(i) Ecosystem maintains biological and non-biological interrelation.

(ii) The overall equilibrium in the ecosystem is important for the well being of all the components.

(iii) The balance of nitrogen, phosphorous, climatic conditions, etc is by proper functioning of ecosystem.

(iv) Energy and nutrients pass through the ecosystem (Food chain).

(v) Ecosystem keeps a watch on environment. Any deleterious changes in the environment, induces adaptations.

Biotic Components of Ecosystems

1. Producers (Autotrophs)

They look for sun for day to day life. They uses chlorophyll and synthesis its required food through photosynthesis using carbon dioxide and water. Eg. Green plants, algae, etc.

2. Consumers (Heterotrophs)

They feed directly on the producers. Based on the food habit they are further classified into different types.

a) Primary Consumers (Herbivors)

It includes grazing animals. Rabbit, cow, deer, giraffe, insects, elephant, etc.

b) Secondary Consumers (Carnivors)

The organisms feed on the Primary **consumers. Snake, cat, fox, wolf, etc.**

c) Tertiary Consumers

The higher-level consumers feed on secondary consumers. They even take primary consumers as food. Eg. Lion, tiger, shark, whales, etc.

d) Omnivors

These are animals which feed on both animals and plants. Eg, human, rat, etc.

e) Deritivores

The animals feed on living and dead maters. Eg. Ants, earthworm, crow, beetles, etc.

Decomposers (Saprophites)

The organisms live in and obtain energy from dead organic material. They includes bacteria and fungous live on organic matter.

Micro Components of Ecosystem

Intangible components required for the proper function of ecosystem. Eg Temperature, light, humidity, pressure, etc.

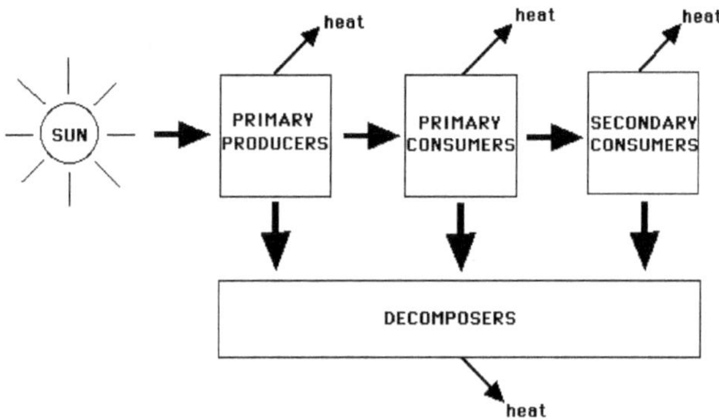

Energy Flow in Ecosystem

Ecosystems are structured with a series of biotic components that are linked together. Energy flows from producer to consumer. Some of it will be lost from the system, as respiration (heat energy) and excreta products. The average efficiency of transfer from plants to herbivores is about 10 per cent. As a result of the loss of energy at each transfer between tropic levels, ecosystems are usually limited to three or four tropic levels.

- The sun is the ultimate source of energy for most ecosystems.
- The primary producers capture a fraction of energy in sunlight and convert it into chemical energy (carbohydrate) that is stored in tissues.
- The energy in tissues is transferred to consumers. Only about 90-95% of energyis transferred to next tropic level.
- The important point is that energy does not cycle through ecosystems. So each energy is taken from sun by producers.

The rate of energy capture is called the **primary productivity** of the ecosystem.

Ecosystem	Primary Productivity (g C/m²/year)
tropical rainforest	2200
grassland	600
deserts	90
coral reefs	2500
lakes and streams	250
open ocean	125

Food Chains

Energy is neither created nor destroyed. In biological systems energy is transferred   from one tropic level to another by the process called eating.

Tropic Levels

Discrete feeding stages in the ecosystem are known as tropic levels. Energy flows through the tropic levels. Producers constitute one tropic level. Primary consumers constitutes another tropic level and so on.

Why there are more herbivores than carnivores?

Energy transfer efficiency is very low (only abour 5-10%. So many primary consumer is required to satisfy the energy needs of few secondary consumers and so on.

Types of Food Chain

- Grazing Food Chain.

- Parasitic Food Chain.

- Detritus Food Chain.

1. **Grazing Food Chain:** Food Chain begins from a green plant.

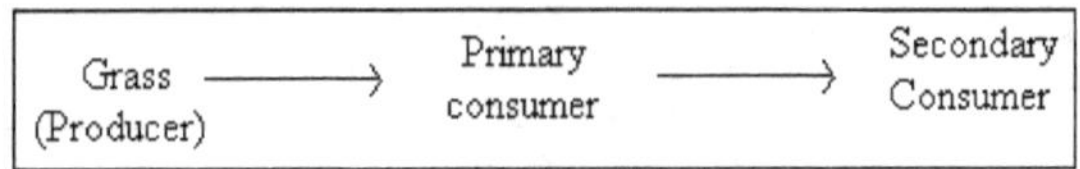

2. **Parasitic Food Chain:** The food chain begins with dead plant and animals.

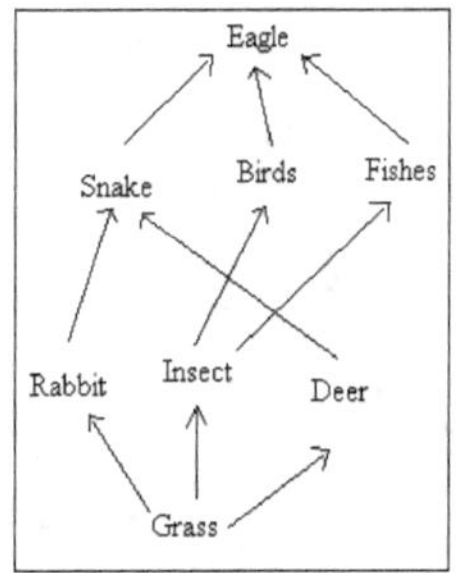

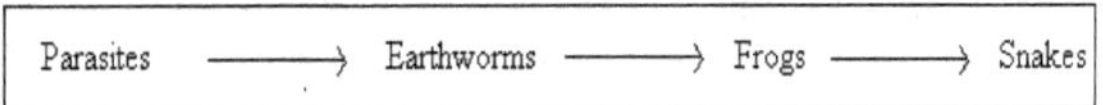

3. **Detritus Food Chain:** The food chain starts from organic matter of decaying animal and plant to microorganisms.

The example is mangrove ecosystem. The food chain follows the following steps.

(a) Mangrove leaves fall into the warm and shallow water.

(b) The leaves are acted by fungi, bacteria, protozoa, etc. living on algae. Algae are eaten by small animals, like crabs, copepods, insect larvae, etc.

(c) Small animals are eaten by small fish and then by large fish followed by fish eating birds.

Food Webs

In actual condition each animal depends on many other animals and/or plants for the food requirements. The diagrammatic representation of such condition form a complex web known as food web.

Difference between Food Chain and Food Web

Sl. No	Food Chain	Food Web
1	It is linear with distinct tropic levels	Non-linear and no clear levels for classification.
2	Linear energy and nutrient flow	Complex energy and nutrient flow
3	Each tropic level depends on one type of food resource.	Depends more than one type of food resource.

Significance of Food Chain and Food Web

- Energy and nutrient flows through food chain and food web.

- It maintains the equilibrium of different animal population in different tropic levels.

- Due to the operation of food chain and food web concentration of environmental toxins in lower tropic level gets increased in the higher tropic levels. The phenomenon is known as bio magnification.

A change in the size of one population in a food chain will affect other populations. This interdependence of the populations within a food chain helps to maintain the balance of plant and animal populations within a community. For example, when there are too many giraffes; there will be insufficient trees and shrubs for all of them to eat. Many giraffes will starve and die. Less giraffes means less reproduction, less food is available for the lions to eat and some lions will starve to death. When there are fewer lions, the giraffe population will increase.

Biological Pyramid

A graph representing the food chain is known as biological pyramid. The graph looks like a pyramid since the energy loss from one tropic level to higher tropic levels. Charles Elton first developed the concept of ecological pyramids.

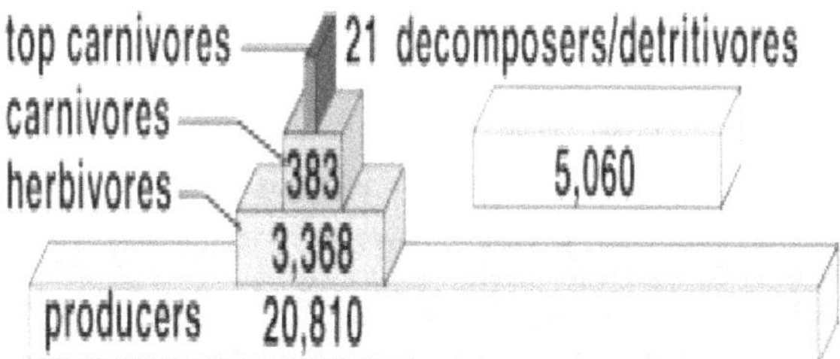

Ecological Pyramids

The graphical representation of relationship between communities in the ecosystem is known as ecological pyramid.

Types of Ecological Pyramids

There are three types of pyramids based on three different kinds of information.

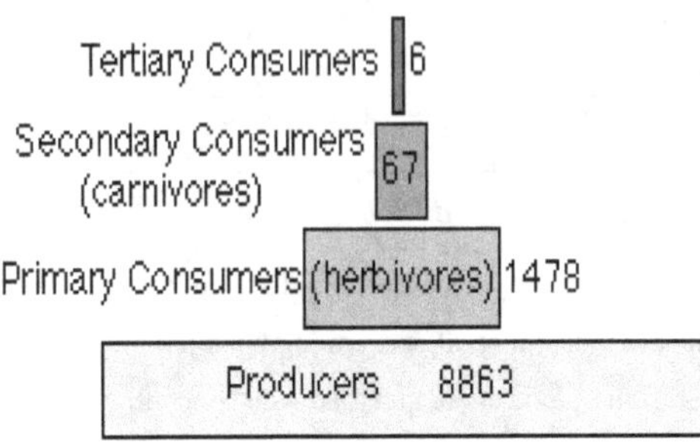

Pyramid of Numbers

The numbers of individuals in each tropic level gives a graph of pyramid of numbers. Small animals are more numerous than larger ones. This graph shows the pyramid of numbers resulting when a census of the populations of autotrophs, herbivores, and two levels of carnivores was taken on an acre of grassland. .

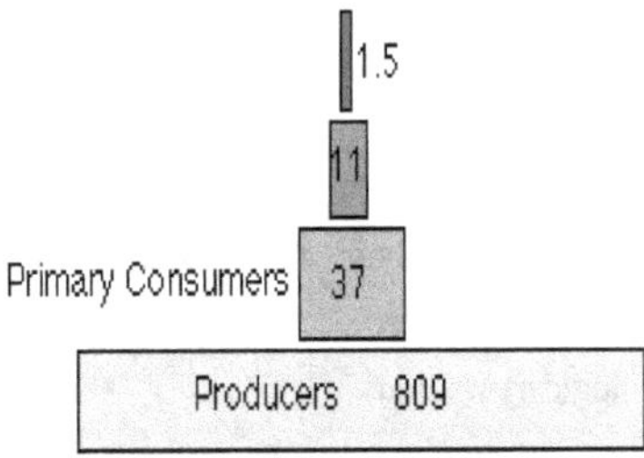

Pyramid of Biomass

The weight of living material in each tropic level is plotted in pyramid of biomass. Since all organisms are made of roughly the same organic molecules in similar proportions, a measure of their dry weight is a rough measure of the energy they contain.

Pyramid of Energy

It is a plot of the energy content of each tropic level. Conversions efficiencies are always much less than 100%. At each link in a food chain, a substantial portion of the sun's energy originally trapped by a photosynthesizing autotroph, is dissipated back to the environment (ultimately as heat).

For example, the total amount of energy in a population of toads must necessarily be far less than that in the insects on which they feed. The insects, in turn, have only a fraction of the energy stored in the plants on which they feed.

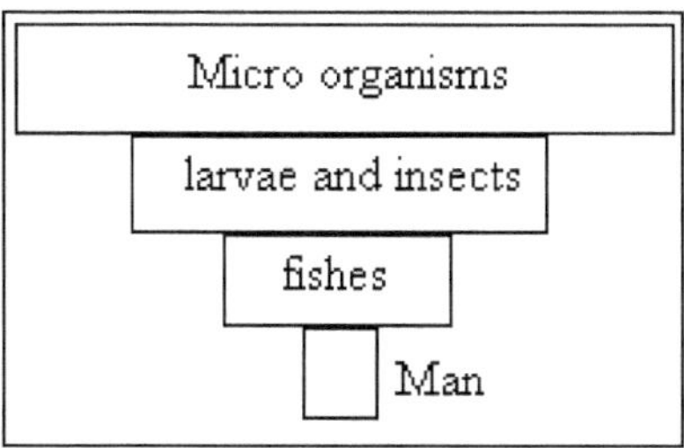

Inverted Pyramid

The shape of the pyramid is inverted when plotted the ecological parameter in the graph. Eg. Number of micro organism is higher than the animals feeding on them.

Energy Flow in Ecosystem's Photosynthesis

Photosynthesis (or phototrophism) is the process by which light energy from the sun is absorbed by plants, blue-green algae and certain bacteria. It is then used to produce a new plant cell material, which forms the food source for plant eating animals (herbivores).

Plants, through the process of photosynthesis, converts light energy and inorganic substances (carbon dioxide, water and various mineral nutrients) into organic (carbon based) molecules. They must contain chlorophyll, which is able to absorb energy from sunlight

$$6CO_2 + 12 H_2O \longrightarrow C_6H_{12}O_6 + 6O_2$$

carbon dioxide   water   glucose   oxygen

Levels of photosynthesis vary considerably. As photosynthesis is a multi-stage process, involving the interaction of several factors. The limiting factors, which affect the rate of photosynthesis, are temperature, light intensity, carbon dioxide concentration and the availability of water.

Biome (Habitat)

It is the environment in which life form has its interaction. It is the natural environment in which an organism lives, or the physical environment that surrounds (influences and is utilized by) a species population. Some of the typical forms of biome are listed below.

Arctic Tundra: This biome has long cold winters and short cool summers.

Deciduous Forest: Mid-latitude deciduous forests have both a warm and a cold season. Much of the human population lives in this biome. "Deciduous" means to fall off, or shed, seasonally. Just as the name implies, these deciduous trees shed their leaves each fall.

Desert: The defining characteristic of a desert is that it is dry. Deserts can be either hot such as the Australian Desert or cold such as the Gobi Desert.

Tropical Rainforest: The tropical rainforest is a hot, moist biome found near Earth's equator. The combination of constant warmth and abundant moisture makes the tropical rainforest a suitable environment for many plants and animals.

Tropical Savannah: The tropical savanna is a biome characterized by tall grasses and occasional trees. Savannas exist in areas where there is a 6 to 8 month wet summer season and a dry winter season.

Taiga: The taiga biome is found in the northern hemisphere close to the polar region. This is cold biome. Winters are long and cold, and the summers are short and cool.

Ecological Niche

It describes the relative position of a species or population in an ecosystem. It helps in understanding the interaction of animal with ecosystem.

Introduction, Type, Characteristic Features, Structure and Functions of Ecosystems

Based on the ecological characteristics ecosystem is classified into different types as in the chart. Ecosystem is self sufficient in its energy and matter. It has independent existence.

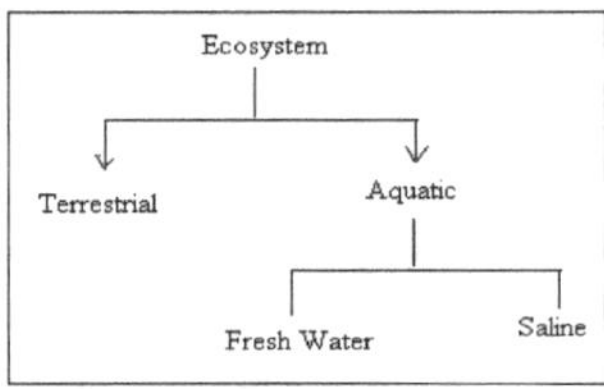

Terrestrial Ecosystem: It is the ecosystem existing on land surface. It includes,

- Forest Ecosystem.
- Grassland Ecosystem.
- Desert Ecosystem.

Forest Ecosystem

Forests are surface on the earth with rich in vegetation and species diversity. A forest is a collection of biological organisms and non-biological factors. From an ecological perspective, the definition of a forest includes all these things, from the trees to the bacteria, and the soil type to the microclimates.

Type of Forests

Tropical Forests: Very dense multistoried forests having diverse trees, shrubs, etc constitute tropical forests. It receives low to high rainfall. Accordingly the forest may be dense ever-green to semi-evergreen.

Mountain Forests: They are confined to mountain areas like Nilgiri hills, Himalayas upto 1500 m altitude. The common trees grow in these type of forests are Eugenis, Piper, Canthium, etc.

Structure of Forest Ecosystem

Abiotic Components	Biotic Components		Decomposers
	Producers	Consumers	
Organic and Inorganic substances present in the soil and atmosphere. The climate, temperature, light, rainfall etc	Mainly trees and the shrubs living on the ground.	**Primary Consumers** Ants, flies, leafhoppers, snails, spiders, etc. (herbivores) **Secondary Consumers** Snakes, birds, mongooses, etc. (Carnivors) **Tertiary Consumers** Lion, tiger, etc.	Microoranisms like fungi, bacteria, and actinomycetis (species of streptomices)

Temperate Forests: This type of forests are characteristics above 1600 m altitude. The trees like oaks, Jasminus, Conifers, etc are abundant.

Alpine Forests: It is high mountain forest above 3000 m altitude. Alpine trees are Sedum, Primulla etc.

Functions of Forest

(i) Raw materials for industry.

(ii) Food resource.

(iii) Fire wood and other essential commodities.

(iv) Minerals and oils.

(v) Climate maintenance and rainfall.

(vi) Rich in biodiversity and supports many living forms.

Grassland Ecosystem

It is a greeny cover on the earth's surface mainly with grasses. Grassland occupies about 19% of earth's surface.

It has large biological productivity.

Characteristics of a Grassland

- Dominated by grass but sometimes with few trees and shrubs.
- Average and erratic rainfall.
- Primary production of food occurs in grassland is consumed by primary consumers.

Structure of Grassland Ecosystem

Abiotic Components	Biotic Components		Decomposers
	Producers	Consumers	
Nutrients in the soil and aerial environment, includes C, H, O, N, P, S, water, CO_2, nitrates, etc.	Mainly grasses and few forbs and shrubs.	**Primary Consumers** Cows, buffaloes, rabbits, sheeps, etc. **Secondary Consumers** Snakes, lizards, birds, foxes, etc **Tertiary Consumers** Hawks which feeds on carnivors.	Microbes like fungi, bacteria and actinomycetes, etc.

Functions of Grassland Ecosystem

- Grassland traps solar energy and the biomass is consumed by primary producers.
- It serves as pool of energy for grassing animals.
- Grasses prevent soil erosion.
- Many insects living and breeding takes place in grassland.
- Many insects reproduce and larvae lives in grasslands.

Desert Ecosystem

A biome with an average annual rainfall of 10 inches or less and sparse vegetation, typically having thin, dry, and crumbly soil. Desert constitutes about 17% in the world.

Characteristics of Desert

- Annual precipitation is less than 25 cm.
- It has poor species diversity.
- The atmosphere of a desert is very dry.
- It generally contains draught resistant, thorned plants, and thick skinned animals.

Structure of Desert Ecosystem

Abiotic Components	Biotic Components		Decomposers
	Producers	Consumers	
Dry soil with scare water and rain. Temperature is either too low or high.	Shrubs, some grasses and few trees.	**Primary Consumers** Insects, reptiles, etc. **Secondary Consumers** Collared lizards, darking beetle, etc. **Tertiary Consumers** Red tailed hawk, agave, gambles's quail	Thermophilic fungi and bacteria.

Functions of Desert Ecosystem

- Desert land is rich in nutrients since not may plants live.
- Store house of micro and macro nutrients.
- Deserts are less in water hence cannot be easily accessible for over exploitation.
- Highly a place for evolution of adoptive animals like camel.

Aquatic Ecosystem – Fresh Water Ecosystem

1. Pond Ecosystem

Pond is a fresh water ecosystem spread over limited area. It is self-sufficient, self regulating ecosystem.

Characteristics of Ponds

- Fresh water stagnant in nature.
- Sometimes ponds may temporary, only seasonal.
- Ponds are more pollution due limited amount of water and over utilization by community.

Structure of Pond Ecosystem

Abiotic Components	Biotic Components		Decomposers
	Producers	Consumers	
Temperature, light, pH, water, Organic and Inorganic compounds	Rooted plants (macrophytes) and floating or suspended lower plants (phytoplankton).	**Primary Consumers** Animals feeding macrophytes, birds feeding phytoplankton (Benthos), etc and zooplanktons,. **Secondary Consumers** Insects, fishes, etc. **Tertiary Consumers** Large fishes, etc.	Bacteria, actinomycetes, fungi, etc.

Functions of Pond Ecosystem

- The fresh water serves as a resource for small water requirements and village forming.
- It contains small algae, plants, animals and self sufficient ecosystem.

- It is the place for community human activity like washing cloths, swimming, drinking water supply, etc.
- Nowadays pond is used for local pollution by small scale, cottage industries generally goes beyond the self-purification capacity of the pond.
- Food resource for small community.

2. Lake Ecosystem

It is a large surface on earth with large amount of stagnant water. In general lake has fresh water with few exceptions.

Structure of Lake Ecosystem

Abiotic Components	Biotic Components		Decomposers
	Producers	Consumers	
Temperature, light , pH, water, Organic and Inorganic compounds.	Phytoplankton, Algae, rooted plants, floating plants.	**Primary Consumers** Zooplanktons, small plant eating animals and fishes. **Secondary Consumers** Small fishes, snails **Tertiary Consumers** Big fishes, fish hunting birds	Bacteria, Actenomycetes, fungi, etc.

Characteristics of Lake Ecosystem

- It is a shallow fresh water body.
- It is a permanent water body with large water resource.
- It helps in irrigation and village community life.
- It is a self sufficient ecosystem.

Functions of Lake Ecosystem

- It is a reservoir for large fresh water.
- It stores water during rainy season and later used for irrigation and community development.
- It is a self sufficient ecosystem.
- Helps in irrigation in surrounding area.
- It is also used for transport of goods and people in the offshore people.
- Provides food resource to surrounding people.

Stream Ecosystem

Stream is small dynamic water body ecosystem. This is a fresh water ecosystem.

Structure of Stream Ecosystem

Abiotic Components	Biotic Components		Decomposers
	Producers	Consumers	
Temperature, light , pH, water, Organic and Inorganic compounds	Green algae, diatoms, aquatic mosses, etc.	**Primary Consumers** Floating algae, phytoplankton. **Secondary Consumers** Small fishes feeding on algae, zooplankton **Tertiary Consumers** Large fishes, fish hunting animals.	Bacteria, Actenomycetes, Fungi, etc.

Characteristics of Steam Ecosystem

- Fresh water dynamic ecosystem.
- Due to mixing of water oxygen concentration may be more.
- Streams have high self purification capacity.
- Streams might be dry during summer.

Function of Stream Ecosystem

- Streams provide fresh water for large community of people living near and around.
- Support small surface agriculture
- Helps for few villages for the community life.
- Streams are used for irrigation of small land area.

3. River Ecosystem

It is a large dynamic fresh water usually originated from mountain or hilly area. In general rivers end in sea or large lake.

Characteristics of River Water System

- It is fresh water free flowing water system.
- It flows through a large area of length and hence provides support to the surrounding areas.
- Rivers might change the place of flow and also deposits large amount of nutrients soil on the side of it.

Structure of River Ecosystem

Abiotic Components	Biotic Components		Decomposers
	Producers	Consumers	
Temperature, light, pH, water, Organic and Inorganic compounds.	Green algae like Cladophora, diatoms, and aquatic mosses.	**Primary Consumers** Fishes and animals with suckers, larvae, etc. **Secondary Consumers** Fish, and other small animals. **Tertiary Consumers** Large fishes, Fish hunting animals.	Bacteria, Actenomycetes, fungi, etc.

Functions of River Ecosystem

- It serves a fresh water resource for large community of people and animals.
- It has high self purification capability.
- It is used for irrigation for large area and length.
- It is used to transport goods and people.
- Generally river is used for dumping solid waste and industrial effluents.

Ocean (Marine) Ecosystem

It is a large saline water body accommodating rich diverse animals and plants. Oceans in the world covers about 71% of the earth's surface. The major are Atlantic, Pacific, Indian, Artic and Antarctic oceans.

Characteristics of Marine Ecosystem

- It is a saline water.
- It occupies a large surface area.
- Vessels like ship, submarines can sail in ocean.
- It is rich in biodiversity.

Structure of Marine Ecosystem

Abiotic Components	Biotic Components		Decomposers
	Producers	Consumers	
About 27% (0.5 M) NaCl, remaining Mg, Ca and K salts, Temperature, light, etc.	Diatoms, diaoflagellates, microscopic algae, etc.	**Primary Consumers** Crustaceans, mollusks, fish, etc. **Secondary Consumers** Herrings, Shad, Mackerel, etc. **Tertiary Consumers** Cod, Haddocks, etc.	Bacteria and fungi, etc.

Functions of Marine Ecosystem

- It serves food for many marine and terrestrial species.
- Provides huge quantity of marine products like drugs, stones, etc.
- Transport and trade.
- Dumping pollutants.
- It is a reservoir for NaCl, and many other minerals.
- It plays a major role in climate change and biogeochemical cycle.

Estuarine Ecosystem

An estuary is a semi closed coastal body of water that has a free connection with sea. River mouths, coastal bays, tidal marshes etc. constitute a estuarine ecosystem.

Characteristics of Estuarine Ecosystem

- It is a transition zone strongly affected by tides of the sea.
- It is dynamic and periodically changes the water characteristics.
- The organisms living in estuarine ecosystems have wide tolerance.

Structure of Estuarine Ecosystem

Abiotic Components	Biotic Components		Decomposers
	Producers	Consumers	
Salts of sodium, potassium, nutrients from river, temperature, pH, etc.	Marsh grasses, sea weeds, benthic algae and phytoplankton.	**Primary Consumers** Oysters, crabs, shrimps, fishes etc. **Secondary Consumers** Sea birds, small fishes. **Tertiary Consumers** Migratory species like eels, salomon.	Bacteria and fungi

Functions of Estuarine Ecosystem

- It is a highly protective ecosystem.
- It is used usually for prone culture and other activity.
- Oesticulture and prone culture carried out in this region.
- Due to high food potential it is used for growing and studying genetically modified species.

Ecological Succession

"Ecological succession" is the observed process of change in the species structure of an ecological community over time.

Ecological succession is a fundamental concept in ecology which states that there is a definable sequence of successional stages through which an ecosystem will pass, before reaching its climax.

Characteristics of Ecological Succession

- It occurs in stages with competition.
- Every ecosystem is developed by the process of ecological succession.
- It is takes place over many deacades before reaching stable state of ecosystem.
- It is a dynamic process and reaching a stable state of community interactions.

Reason for Ecological Succession

- Every species has a set of environmental conditions under which it will grow and reproduce most optimally.
- As long as the ecosystem's set of environmental conditions remains constant, those species optimally adapted to those conditions will flourish.
- The original environment may have been optimal for the first species of plant or animal, but the newly altered environment is often optimal for some other species of plant or animal.
- Succession is one of the major themes of our Nature Trail.

Example

In a sun-loving pine tree area invasion of hardwood trees (including ash, poplar and oak) leads to competition. The fast growing hardwood trees give shades to pine trees. Thus pine trees can not grow comfortably under the hardwood trees. This leads to decline in pine trees. Ultimately the entire area is occupied by hardwood trees at the end of succession.

Classification of Succession

(a) Autotrophic Succession

It is characterized by early dominance of autotrophic (green) life forms like many plants.

(b) Heterotrophic Succession

In this type of succession the early dominance of heterotrophs like bacteria, acteinomycetes, fungi and animals are observed.

Stages of Ecological Succession

(i) **Nudation:** It is the presence of bare area. Only abiotic components will be available.

(ii) **Invasion:** It is the arrival of life forms. Plants first invade in the area called as pioneer community.

(iii)**Competition:** In this stage many communities are established and they compete each other for space, food, etc.

(iv)**Reaction:** Different organisms modify to suit the environment. The favourable communities are formed which are called seral communities.

(v) **Stabilization:** It is the final stage in which a stable community is established. It is called climax community. Such community exists for a long period of time in the typical ecosystem formed.

End of Succession: There is a concept in ecological succession called the "climax" community. The climax community represents a stable end product of the successional sequence.

Introduction to Biodiversity

It refers to the variety and variation of life on a given area such as ecosystem. It includes the diversity of genetic materials within species, the variety of species in all taxonomic groups, and the array of communities, ecosystems, and landscapes within which species evolve and coexist. Biological diversity is the variety and variability among living organisms and the complexity among them.

Importance of Biodiversity

- It provides the living environment and associations with different life forms.
- It gives and food and other resources like fiber, timber, medicines, fuel, etc. essential for comfortable life.
- Proper function of food chain food web and nutrient cycle occurs through biodiversity.
- It gives raw material for many industrial processes.

Biodiversity is inspirational: We rely on biodiversity for enjoyment and recreation, spiritual fulfillment and for our cultural heritage.

- The major drugs produced from plants has a current market value of about $ 200 million per year.
- Out of top 150 prescription drugs in US there 118 comes from plants (74%), fungi (18%), bacteria (5%) and vertebrates (3%).
- Over 80% of world population depend on traditional plants for medicine.
- Loosing a tree species means ultimate loss of 3-4 potentially valuable drug, which cost $ 600 million per year.

Characteristics of Biodiversity

- It is characterized by large number of living species living together and interacting well in a stable manner.
- A stable population and food chain.
- Imbalance in biodiversity leads to species extinction and loss of biodiversity.

Classification of Biodiversity

Genetic diversity : It is variation within individual species. It arise due to smaller variation in genetic material. It is responsible for the different traits in species. Eg. Different dog varieties, different rice varieties, etc.

Species diversity: It is the existence of many species in an ecological area. It is characterized by number of species of animals and plants interact in an area as a stable community.

Ecosystem diversity: Ecosystem diversity means the existence of many type of ecosystem in a given area and many life forms interact in that environment. For example in a typical ecosystem, water body associated with dense plants and in continuum there is grassland can be considered as ecosystem diversity.

Value of Biodiversity

It is the process of assigning importance to biodiversity with respect to different perceptions of human being. Three main approaches have been used for determining the value of biological resources:

- Assessing the value of nature's products-such as firewood, fodder, and game meat--that are consumed directly, without passing through a market ("consumptive use value");
- Assessing the value of products that are commercially harvested, such as timber, fish, game meat sold in a market, ivory, and medicinal plants ("productive use value"); and

- Assessing indirect values of ecosystem functions, such as watershed protection, photosynthesis, regulation of climate, and production of soil ("non-consumptive use value"), along with the intangible values of keeping options open for the future ("option value") and simply knowing that certain species exist ("existence value").

Consumptive Value: Consumptive value refers to non-market value of resources such as firewood, game, meat, etc. Such resources are consumed directly, without passing through a market.

Product	Source	Use
Penicillin	Fungus	Antibiotic
Streptomycin	Actinomycete	Antibiotic
Tetracycline	Bacterium	Antibiotic
Quinine	Cinchona Bark	Antimalarial
Taxol	Taxus	Anticancer
Resepine	Rauwalfia	Hypertension
Morphine	Poppy	Analgestic

Productive Value: Productive value refers to the commercial value of products that are commercially harvested for exchange in formal markets, such as game meat, timber, fish, ivory, medicinal plants. They are included in national income accounts like the GNP. This includes fruits such as Brasil nuts, and latex from rubber-tapping).

A single corn plant can transfer 60 gallons of water from soil to atmosphere. In analogy, a rainforest tree can transfer about 2.5 million gallons of air from soil.

Social Value: "Biodiversity provides a huge package of variety, which the human mind needs to be fully functional". Biodiversity is a direct source of pleasure and satisfaction. It contributes to quality of life, outdoor recreation and scenic enjoyment. There are many plants and animals are given more value in the society and given religious attachment to it., Eg. Tulsi, lotus, cow, monkey, etc.

Ethical Value: Ethics is inherent duty to respect the nature beyond the supremacy of humans. It is the duty based on higher moral principles and compassion towards all living forms in the plant. A nation's biodiversity is also reflection of its ethical principles on natural life forms. India's rich biodiversity could be due to its ethical perspective right from Vedic period. As we aware, plants and animals are praised with Gods and given due credentials. Animal ethics is incorporated in many poetries and writings in Indian literature.

Aesthetic Value: Biodiversity has aesthetic for humans, both in the form of specific taxa such as flowers, birds, trees and as components of natural or semi-natural landscapes. A sense of satisfaction among population when they aware rare plants and animals are within the coutry they belong. The recent concept of ecotourism is based on aesthetic value by encouraging tourists in enjoying the biodiversity in the natural environment. National parks, snake parlours, horticultural species, etc constitute aesthetic value.

Opinion Value: It is based on individual's opinion either based on culture, religion, etc. It is the respect to peoples feeling. For some strata cow has in their option higher value than a horse. For few other sheep may have better placed opinion and so on.

Biodiversity at Global, National and Local Levels

The number of species globally may be 13.6 millions. In this only 1.76 million has been discovered and search for the remaining is on the way. About seven per cent of the world's total land area is home to half of the world's species, with the tropics alone accounting for 5 million.

Group	Number of Species in India	Number of Species in World
Mammals	350	4629
Birds	1224	9702
Reptiles	408	6550
Amphibians	197	4522
Fishes	2546	21,730
Flowering Plants	15,000	250,000

In the national level India there are 12% of global flowering plants and about 7% of total animal species in the world.

India has over 2000 species of medicinal plants. She has two biodiversity hot spots with high endemism and endangered species.

Further rich varieties of flowering plants agricultural varieties of crops constitute a biological diversity. Due to diverse and rich life forms India is considered as megadiversity nation.

Local Biodiversity

In local level biodiversity is better classified into four classes based on species richness and distribution.

    (i) Point Richness: It refers to richness of species in a single point of specified area.

    (ii) Alpha (α) Richness: It is classification based upon the number of species present in a small homogenous area.

    (iii) Beta (β) Area: It is based on the rate of change of species across different habitat.

    (iv) Gamma (γ) Area: It is identified based on the rate of change of species in a large landscape area.

Threats to Biodiversity

Biodiversity is nature's gift. It is the result of millions of years of evolution and ecological succession. Such stabilized communities to be preserved.

In recent decades maintaining biodiversity in a given area of is a great task. This is because the activities in the name of industrialization and economic revolution many habitat being destroyed or modified. It affects the comfortable life of many living forms. Thus it is a great threat to biodiversity. Most threats to biodiversity are directly or indirectly attributable to a growing human population.

Human activities are endangering other species around the globe. Extinction is part of the evolutionary process, but today's rate of extinction is much greater than the scale at which species disappear due to evolution alone. Species are now vanishing faster than at any other time in Earth's history

Even conservative figures predict a loss of at least one per cent of existing species per decade, which means we're losing at least two species an hour.

Reason for Loss of Biodiversity

- The over-use or over-harvesting of plants, animals or natural resources threatens Earth's biodiversity. Over-exploitation, such as logging, hunting or fishing can reduce species numbers to the brink of extinction.
- It is estimated that global warming has the potential to destroy 35% of the world's existing terrestrial habitat. Climate change may cause local species loss as high as 20% in the most vulnerable arctic and montane habitats.
- Deforestation, desertification, drought, famine, etc leads to loss of biodiversity.
- Illegal trade, smuggling and biopiracy.

- Most threats to biodiversity are directly or indirectly attributable to a growing human population leading.

- Pollution
- Over-exploitation
- Climate change
- Invasive Species
- Habitat loss

Red Data Book

It is a catalogue containing name of plant and animals which are threatened species. India lists 44 plants in Red Data book as critically endangered, 54 endangered and 143 as vulnerable. She ranks 2nd in terms of number of threatened mammals and 6th among countries with the most threatened birds.

In India botanical survey of India compiled three volumes of Red Data book having endangered plant species. The list of rare plants is given in Green Book. There is a Blue Book which has the UNEP compiled list of endangered species.

Habitat Loss

The primary threat to the world's biodiversity is habitat destruction, which alters or eliminates the conditions needed for plants and animals to survive.

Habitat is the sum of the total of the environmental factors, food, water and shelter which is needed for a given species to live comfortable and reproduce in a given area.

Habitat can be destroyed or degraded by two basic ways.

(i) Quantitative loss is loss in a particular area which no longer present. For example if a wetland is lost, it is a quantitative loss.

(ii) Qualitative changes involve the degradation of structure, function or composition of habitat. Poisoning of water by chemicals is a qualitative loss.

Habitat loss is in general irreversible, can not be restored. Logging of tropical rain forests, urbanization etc are irreversible habitat loss.

Causes of Habitat Loss

- Deforestation.
- Natural calamities.
- Industrialization and Urbanization.
- Mining and associated activities.
- Population and cultural changes.

Control of Habitat Loss

- Reforestation.
- Implementation of sustainable development concepts
- Better technology and optimal uses of resources.
- Alternate technologies which decreases the impact on habitat loss.
- Protected, designated area should be prevented from economic exploitation.

Poaching of Wildlife

Poaching is illegal killing of animals by man for meat, skin, sport, organ transplant, research, etc. It is more of a commercial activity. Poaching leads to loss of biodiversity. This is because decrease in population in one species affects many other species.

The worst hit species are tigers, lions, elephants, beat, etc. Poaching also fuelled by international market demands illegal trade. For example global demand of ivory is the cause of decrease in elephant population in the world. In 1977 the African population of elephants come down from 1.3 million to 600,000.

Poaching of some animals like peacock, monkey, tortoise, Tiger, etc also for medicinal purpose.

Case Study

Polar Bear Poaching

Polar Bears are endangered species and more than half is found in northeastern Russia. The population is declining fast due to illegal hunting. In 1956, Russia made polar beat hunting illegal and made bilateral treaty with other states namely, Canada, Denmark, Greenland, Norway and United States. Although it is protected by various laws, illegal poaching is continuing in remote areas and very little has been done to prevent this due to the collapse of Russian economy.

Control of Poaching

   (i)   Enforce laws strictly.

   (ii) Trade related to animal and its products to be banned.

   (iii) Tourism based on animal game, gambling, etc should be discouraged.

Man Wildlife Conflicts

Man maintains his supremacy and tries to conquer others, including nature and animals. One of the wildlife, the man also needs to be protected though humpty number of laws made himself for him.

Man takes any small cause for his favour. If an elephant destroyed certain area of agriculture and his next thing to do is compensation by killing the elephant by any crooked means. It is the fact that the land area was belong to the elephant habitat till before it underwent deforestation.

Historically hunting is a game and was done by Royal king to show his Majesty and power and is being spread across all sections.

As a case study, few years back a tiger took away a child from tourist in Bannargatta National Park, Bangalore. In Sambalpur, Orissa about 195 people were killed and retaliation 98 elephants were killed and 30 injured. In India on average, more than 15 people are killed every year. Human settlement has interfered with dispersal patterns of wild animals leading to animals invading farmlands, destroying crops and killing people.

Causes of Conflicts

(a) Deforestation, urbanization, etc.

(b) Captive animals in parks, restricted area are unhappy and angry.

(c) Self protection due to anticipation of thread on both the sides or any one side.

(d) Avoiding the destruction of commodities, agriculture, etc.

(e) Accidental conflict when animals invade in residential area.

Control Measures

(i) Not invade in the privacy of animals and not occupy illegally the forest and protected area.

(ii) Fencing the agricultural area.

(iii) Avoid settlement in the near forest area.

(iv) More area should be allotted to animals and adequate supply of food, water, etc. shall be maintained in their area.

(v) As for as possible avoid entering wildlife area so that they are less worried about their habitat protection.

(vi) Provisions of Wild life protection laws to be used strictly by human beings on human beings only.

Endangered and Endemic Species in India

A species is in great danger by drastically coming down its population is said to be endangered (threatened) species. It is in urgent need in protection from being extinction. India has 210 species of animals considered to threatened and it accounts about 2.9% of worlds threatened species. In addition as many as 3,000 to 4,000 higher plants has been considered as threatened.

Endangered Animals-Peacock, tortoise, red fox, tiger, lion, etc.

Endangered Plants: Sandal wood tree, teakwood tree, Aloe vera, *Cycas beddomei, Pterocarpus santalinus,*

The species confined to a particular region are known as endemic species. About 33% of flowering plants, 62% of amphibians, and 50% lizards are endemic to Indian sub-continent.

Endemic Animals: Vivparous toad, Indian salamander, reticulated python, etc.

Endemic Plants: *Indigofera uniflora, Osbeckia aspera*, etc.

Extinct Species in India

Extinct species is no longer seen in the recent past and all probable reason the species is no more existing in the world.

Extinct Animals: Dodo, Passenger pigeon, etc.

Extint Plants: ilex khasiana, actinodaphne lanata, etc.

India Megadiversity Nation

A very small number of countries, mainly in the tropics, possess a large fraction of world species diversity. The countries which possess the greatest species richness are recognized as mega diversity countries. It attracts special international attention.

India is one of the 12 megadiversity countries with only 2.4% of the global land. Because of its richness in overall species diversity India is recognized as one of the megadiversity regions of the world.

Present Status of India's Biodiversity

(i)   India has a rich and varied heritage of biodiversity, encompassing a wide spectrum of habitats from tropical rainforests to alpine vegetation and from temperate forests to coastal wetlands.

(ii)   India contributes significantly to this latitudinal biodiversity trend. With a mere 2.4% of the world's area, India accounts for 7.31% of the global faunal total with a faunal species count of 89,451 species.

(iii) India is one of the 12 megadiversity countries with rich biodiversity..

(iv)   She has 45,500 plant species, representing about 12% of the world's biota.

(v)   Of about 1.7 million species globally described and recorded in scientific literature, India has about 1,26,200 species. It ranks tenth in the world both in respect of richness of flowering plants (17,000spp.) and mammals (372).

(vi)   India ranks 7th as far as the number of species contributed to agriculture are concerned. India has been a primary centre of domestication for rice, sugarcane, banana, tea, mangom cucumber, citrus, jute, minor millets, vignas, brassicas, jack fruit, aloeasia, colocasia, cardamom, black pepper, ginger, turmeric, cucurbits, bamboos, etc. and a secondary centre of domestication for sesame, tomato, potato, maize, soyabean, etc.

(vii) The Indian region is one of the most diverse biogeographic regions of the world, embracing a wide range of topography from perpetually snow covered high Himalayan ranges to plains at sea level, low lying swamps and mangroves.

(viii) Human population further contributes to become India a Megadiversity nation. India reached the population of one billion people in 2000, comprising about 16% of the world's population. Presently it is more than 1.2 billion and increasing estimated to be in the rate of 1.92%.

The compounded with high cattle population, estimated to be 450 million, contributes 18% of the cattle population in the world. Thus suitably India is a mega-diversity nation.

Hotspots of Biodiversity

Hot spots are areas that are extremely rich in species, have high endemism, and are under constant threat. Among the 25 hot spots of the world, two are found in India extending into neighbouring countries-the Western Ghats/Sri Lanka and the Indo-Burma region (covering the Eastern Himalayas).

These areas are particularly rich in floral wealth and endemism, not only in flowering plants but also in reptiles, amphibians, swallow-tailed butterflies, and some mammals.

Philippines, one of the "hottest" of the Hotspots, is featured in National Geographic Magazine's July, 2002 edition.

Global Biodiversity Hotspots

Sl. No.	Hot Spot	Sl. No.	Hot Spot
1	Tropical Andes	14	Mediterranean Basin
2	Mesoamerica	15	Caucasus
3	Caribbean	16	Sundaland
4	Brazil's Atlantic Forest	17	Wallacea
5	Choco/Darien/Western Ecuador	18	Phillipines
6	Brazil's Cerrado	19	Indo-Burma
7	Central Chile	20	. South-Central China
8	California Floristic Province	21	Western Gnats/Sri Lanka
9	Madagascar	22	SW Australia
10	Eastern Arc and Coastal Forests of Tanzania/Kenya	23	New Caledonia
11	Western African Forests	24	New Zealand
12	Cape Floristic Province	25	Polynesia/Micronesia
13	Succulent Karoo		

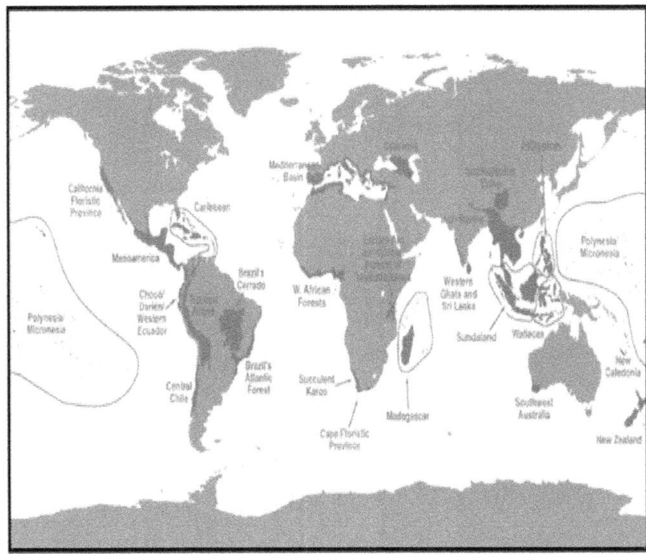

Bio Geographical Classification of India

India has been classified into 10 biogeographical regions

Sl.No	Name	Description	Representative Plants and Animals
1	Trans-Himalayan	This is very cold and arid region of Himalaya between 4500-6000 m. The only vegetation is alpine steppe. Most of the area is bare rock and glacier.	Pine, Wild sheep, Snow Leopard, Blacknecked Crane
2	The Himalayan	It is composed of North-West, West, Central and East Himalayas	Cork tree, Pine. Wild beer, Leopard
3	The Indian desert	It is tropical thorn forests, dry deciduous forests and sandy deserts. Thar desert possesses insects, reptiles and birds. Antelopes and desert fox etc live there.	Acacia, Date Palm, Camel, Wild ass, Desert cat, Fox, Rat.
4	The semi-arid zone(s)	The region consists of west part of Thar, Gulf of Kutch and whole of Kathiawar peninsula.	Acacia, Peepal, Date palm, Gir Lion, Tiger
5	The Western Ghats	The western region of India gets high rainfall. It contains forests with high diversity. The animal species include Nilgiri langur, lion tailed macaque, Malabar grey hornbill, amphibians, etc.	Peepal, Jamuna, Bahera, Tortoise, Frog, Lizards, Snakes.
6	The Deccan Peninsula	It is the land area covering about 43% of India. It is between Western and Eastern Ghats. The total elevation is from 900-300m. A large part is covered by tropical and tropical deciduous forests.	Acacia, Jamuna, Pine, Sloth bear, Tiger, Cheetah, Elephant.
7	The Gangetic Plain	It is the most fertile region of the India. The soil is alluvial deposits of the Ganges and it tributaries.	Sal, Acacia, Mango, Rhinoceros, Gazzel, Alligator, Turtle.
8	The West Coast	There are 26 species of fresh water turtles, tortoises and five species of marine turtles live in costal region.	Coconut, Banana, Cashew nut, Dolphin, Turtle, Alligator, Molluscs
9	The North-East India	The region is rich in evergreen and semi-evergreen rain forests. It contains 390 species of mammalians. The highest population of elephants in India is in North East region.	Bamboo, Jack Fruit, Chestnut, Elephant, Rhinoceros, Deer, Porcupine.
10	The islands of Andaman and Nicobar	It consists of a group of 325 islands only 21 are inhabited. The rainfall in this region is high due to both Northeast and Southwest monsoons. About 2,200 plants are found here.	Bahera, Jack Fruit, Cardamom, Coconut, Dolphin, Alligator, Molluscs.

Conservation of Biodiversity

By considering the rate of extinction of animals and plants there are remedial measures taken to protect the species. The measures are called as conservation strategies. The World Commission on Environment and Development (WCED) constituted by the General Assembly of the United Nations in 1986. It has provided a major boost and endorsement to the need for conserving the world's rich biodiversity particularly that of the tropical areas. In 1992, the 'Earth summit' held under the auspices of the United Nations Conference on Environment and Development (UNCED) at Rio de Janeiro adopted a convention on biological diversity and laid stress on the integration.In addition "Convension on International Trade in Endangered Species, 1975" (CITES) prohibits the trade of 900 species. It further restricts 2900 species since they are endangered. The treaty is ratified by 160 nations in the world including India.

The most widely accepted definition of 'conservation', presented in 1980 in World Conservation Strategy by the International Union for Conservation of Nature and Natural Resources, is that of "the management of human use of the biosphere so that it may yield the greatest sustainable benefit while maintaining its potential to meet the needs and aspirations of future generations."

Introduction to Biodiversity

1. Leads to conservation of essential ecological diversity and life support systems.
2. Preserves the genetic diversity of plants and animals.
3. Ensures the sustainable utilization of life support systems on earth, food, property, shelter, etc.
4. Provides a vast knowledge of potential use to the scientific community.
5. Maintain remaining biotic communities in their natural form.
6. Provides immediate benefits to the society such as recreation and tourism.
7. Provide controls against the changes brought about by other forms of land use.

General Measures for Conservation and Sustainable Use

1. Develop national strategies, plan or programs for the conservation and sustainable use of biological diversity.
2. Integrate as far as possible and as appropriate, the conservation and sustainable use of biological diversity into relevant sectoral or cross-sectoral plans, programs and policies.

Ex-Situ (Outside the Habitat) Conservation

Biological diversity (Both Animal and Plant species) may be conserved outside the areas where they naturally occur. It is known as *ex-situ* conservation. The ex-situ conservation area are zoological gardens, botanical gardens, biological gardens, tissue and cell culture banks, seed banks, etc.

National Bureau of Plant Genetic Resources (NBPGR)

It uses cryo preservation (Low temperature storage) technique to preserve agricultural and horticultural crops. It is located in New Delhi.

National Bureau of Animal Genetic Resources (NBAGR)

It preserves the semen of domesticated bovine animals. It is located at Karnal, Haryana.

National Facility for Plant Tissue Culture Repository (NFPTCR)

It preserves crop plants, trees by tissue culture methods. It is a subsidiary of NBPGR.

Advantages of Ex-Situ Conservation

- Special care and protection is given.
- Animals and plants are protected from enemies and predators.
- Comfortable environment is given for breeding and propagation of species.
- Required food and other amenities are provided and thus need not struggle for existence.

Disadvantages of Ex-Situ Conservation

- The species is kept in artificial environment and feel uncomfortable.
- Free life is not guaranteed and thus cannot enjoy its natural habitat.
- The cost of maintenance and expertise requirement is high.
- All need of the species cannot be satisfied.
- Frequent human activity disturbs the life style.

In-Situ (within the Habitat) Conservation

It is the protection and conservation of animals and plants in its natural habitat. About 4% of the geographical area is to be reserved for in-situ conservation. The following areas may be set aside for *in-situ* conservation:

a) National parks and sanctuaries
b) Biosphere reserves
c) Nature reserves
d) Reserved and protected forests
e) Preservation plots and sample plots
f) Reserved trees.

National Parks and Sanctuaries

These are legally constituted protected areas for conserving both the flora and fauna. In India, the Wildlife (Protection) Act of 1972 empowers the State Governments to declare and areas as a 'Sanctuary or National Park' after following the procedure that has been laid down. India has 80 national parks and 492 wildlife sanctuaries.

In 1936, the first National Park was created in India and named as the Hailey National Park, now known as the Corbett National Park, Uttar Pradesh.

Game Reserves

Under section 36 of the Wild Life (protection) Act, the State Government may declare any area to be game reserve and no hunting of wild animals is permitted in such a reserve except with the prior permission of the chief wildlife warden or an authorized officer.

Biosphere Reserves

Biosphere reserves are areas set aside for conserving the flora, fauna and environment. These reserves are established under the Man and Biosphere Programme of the UNESCO. It covers a large area more than 5000 sq. km. India has 7 biophere reserves. The main features of biosphere reserves are:

- They are representative areas of specific terrestrial and coastal environments of a country, continent or the entire earth that must be conserved for posterity.
- They are representative examples of natural or minimally disturbed ecosystems.
- The extent and size of such areas is large enough to function as a unit of conservation.
- Biosphere reserves remain and function as an open system and changes in land use are not usually allowed.

Preservation Plots, Sample Plots and Protected Trees

Preservation plots, sample plots and protected trees are important for conservation of biological diversity. This process was started in India in 1905. At present there are 309 preservation plots all over the country 187 in natural forests and 122 in plantations. These cover an area of about 8500 hectares. There are 1905 sample plots in different forests types and 537 protected trees of different species in various forests.

Advantages of in-situ Conservation

- It is easy, reliable and convenient method of conservation of species.
- The species is comfortable since it is in its natural habitat.
- The natural life style of the life form is maintained.

Disadvantages of in-situ Conservation

- It is required to preserve the natural habitat by all means.
- Generally large area is to be maintained with administrative problems.
- No actual control over the animals and plants.

Questions

1. Apply the ecosystem concept to man and options available to him for energy flow.

2. What producers, consumers and decomposers and their inter relationship?

3. What is the energy flow efficiency in different ecosystem and explain based on law of conservation of energy?

4. Explain the significance and process of succession.

5. What are food chain and food web and their significance?

6. Explain different types of ecological pyramids.

7. Explain the function and energy flow in

    a) Forest ecosystem.

    b) Grassland ecosystem.

    c) Desert ecosystem.

    d) Aquatic ecosystem.

8. What is biological diversity and explain its significance in ecosystem equilibrium.

9. Explain genetic and ecosystem diversity

10. What is the value of biodiversity? Explain different type of values.

11. "India-a Mega diversity Nation"-Explain.

12. What is a hot spot and measures taken to preserve it?

13. Explain endemic and endangered species.

14. Write note on endangered and endemic species in India.

15. What are the steps taken by Government of India for in-situ and ex-situ conservation?

Unit II

Environmental Pollution

Definition – causes, effects and control measures of: (a) Air pollution (b) Water pollution (c) Soil pollution (d) Marine pollution (e) Noise pollution (f) Thermal pollution (g) Nuclear hazards – soil waste management: causes, effects and control measures of municipal solid wastes – role of an individual in prevention of pollution – pollution case studies – disaster management: floods, earthquake, cyclone and landslides.

Pollution

The word Pollution is derived from Latin word 'pollutionem', meaning 'defile or make dirty. Pollution is defined as the deviation from the natural composition of a part of the environment resulting in adverse effect on life.

Pollution is an unfavorable change in the physical or biological characteristics of our air, land and water (or basic amenities) causing harmful effect on our life or that of other desirable species and cultural assets.

Therefore, it is defined as, "The addition of the constituents to water, air, or land, which adversely alter the natural quality of the environment".

According to US President's Science Advisory Committee, Environmental pollution is the unfavourable alteration of our surroundings, wholly or largely as a by-product of man's actions, through direct or indirect effects of changes in the energy pattern, radiation level, chemical and physical constitution and abundance of organisms.

Natural pollution is the act of nature and man can not control its occurrence that easily. Eg, natural forest fires, volcanic eruptions, earthquake, etc.

Artificial pollution is due to the activities of man and is under his control. Eg, industrial effluents and emissions, thermal and radioactive discharges, etc.

Pollutants

The material, which causes the pollution to the environment, is called pollutants. Pollutants are undesirable substances present in the wrong place, at the wrong time and in the wrong quantity.

Classification of Pollutants

Sl.No.	Classification	Definition	Example
1	**Primary Pollutants**	The substances emitted directly from an identifiable source.	Sulfur dioxide, nitrogen oxides, etc.
2	**Secondary Pollutants**	Derived substances or effects from primary pollutants.	Peroxyacetyl nitrate.
3	**Biodegradable Pollutants**	The pollutants that can be readily decomposed by natural processes.	Municipal sewage.
4	**Non bio degradable Pollutants**	Resistant to biological action present in the environment.	DDT, mercury, lead, etc
5	**Natural Pollution**	Not under human control	Volcanic eruptions, earthquake, etc
6	**Artificial Pollution**	Due to human activity	emission, sewage,

Pollutant is a harmful solid, liquid or gaseous substance present in such a concentration in the environment, which tends to be injurious for the whole living biota.

Pollutants need not be a tangible material substance. Noise, radiation, thermal energy etc are also pollutants. Invariably pollutants are required to cause pollution.

Definition Causes Effects and Control Measure of Air Pollution

Definition of Air Pollution

Engineers Joint Council of USA defines air pollution as "the presence in the outdoor atmosphere of one or more contaminants, such as dust, fumes, gas, mist, odour, smoke or vapour in quantities, with characterists, and of duration much as to be injurious to human, plant or animal life or to property, or which unreasonably interfere with the comfortable enjoyment of life and property".

According to the Bureau of Indian Standards, IS-4167 (1966) air pollution is "the presence in ambient atmospheres of substances, generally resulting from the activity of man, in sufficient concentration, present for a sufficient time and under circumstances such as to interfere with comfort, health or welfare of persons or with reasonable use or enjoyment of property".

Man can survive for 5 weeks without food, 5 days without water and less than 5 minutes without air. Man breathes, on an average, 25000 times a day at rate of 1-2 litres of air per breath. World Environment day is marked on June 5 every year.

Compared to water about 2 litres a day man consumes about 20,000 times more of air by volume and 25 time more by weight. A concentration f more than 0.3 mg/l or 3,00,000 µg/m3 of lead in water is considered harmful to man but a concentration of 1.5 µg/m3 of it in air is deadly harmful.

Pollutants: Thermal power amounts to more than 70 percent of power generation. 1000 tonnes of coal is burnt to produce. 1 MW power, produces the following pollutants (in tones / MW / year)

Ash 750, carbondioxide 7000. Sulfur dioxide 12, Oxides of Nitrogen 20. In addition, carbon monoxide, and cancer causing heavy metals, present as impurities in coal, are also released.

Thermal power stations are responsible for about 14% of the global air pollution, while transport systems and industry contribute 60% and 16% respectively.

Components of Air Pollution

Toxic Gases: Oxides of Nitrogen and Sulphur, (NO, NO2, SO2, etc), Carbon monoxide, Carbon dioxide, etc.

Toxic Sprays: Most of the agricultural sprays like insecticides, herbicides, fungicides, etc.

Radiations: Natural and man made radiations like, cosmic rays, UV radiations, X-Rays, Neutrons, alpha particles, etc.

Thermal Emissions: High temperature emissions to the environment from industries.

Dust: It is due to volcanic eruptions, Hurricane, mining, metallurgical process, etc.

Sl.No.	Causes	Consequences	Control
1	Industrial Emissions, dust	Respiratory and neuro system disorders.	Follow the emission standards. alternative fuels to be used.
2	Mining, and related activities	Digestive system disorders	Vehicles to be in condition, frequent emission tests.
3	Accidents and fires	Corrosion of structures.	Not to use adulterated fuels.
4	Vehicular Emissions	Necrosis and skin problems	Primary treatment of emissions before to environment.
5	Radiations	Green House Effect and climate change.	Proper sitting of industry in less populated area.
6	Smoking and related	Loss of agriculture productivity.	Green forestry to absorb some of the pollutants by plants.
7	Agricultural Sprays	Ozone hole and radiation	Controlled industrial and exploration activities.
8	Volcanic eruptions, forest fires, etc.	Anemia and blood borne diseases	Nuclear accidents must be avoided.
9	Refrigeration and Air conditioning (CFCs)	Acid rain and death of aquatic life forms.	Alternate material for chlorofluorocarbons (CFCs).

Photochemical Smog

$$CH_4 + O_2 + NO_2 \longrightarrow CH_3-\underset{\underset{O-O'}{}}{\overset{\overset{O}{\parallel}}{C}}{\;}^{NO_2}$$

Peroxyacetyl Nitrate (PAN)

The formation of smoke in sunny days in and around industrial area is known as photochemical smog. It is due the reaction of pollutants with oxygen and sunlight.

The peroxyacetyl nitrate mixes with fog and dust forms smoke called photochemical smog.

Los Angeles Smog

It is the smog formed in City Los angels due to vehicular emissions followed by photochemical reactions.

London Smog

It is the smoke formed above the industrial area in London. The smoke was due to sulphur dioxide, dust and water vapour. The smog is also known as sulphur smog.

Indoor Air Pollution

It is pollution in a confined and closed area. Indoor air pollution often leads to severe health hazards like irritation to eyes, nose, throat, headache and respiratory problems. It is proved that indoor pollution increases the tension and decreases the productivity of workers.

Causes of Indoor Air Pollution

Improper ventilation and exhaust, use of synthetic waxes, paints, fumigants etc. In the 'Black Hole of Culcutta', 146 prisoners were imprisoned in a room with two very small windows with adequate supply of oxygen for all. But only 23 survived.

In Mardh 12, 1944 the Salarno train disaster killed people, in a train accident. When the train was traveling in a long tunnel, the oxygen in the tunnel was used up by the passengers in a matter of minutes and 512 died due to suffocation.

The study concludes that the death was due to changes in the physical condition of the air, leading to heat retention because of poor ventilation.

Control of Indoor Air Pollution

Source Removal: Sources other than human such as motor vehicles, wastes, appliances which are not in use etc to be moved out. Smoking must be practically prohibited in indoor.

Improved Building Design: Better design in house to remove combustion production during cooking and related enjoyment activities.

Preferably use low fuming heating agents, attached with chimney in the kitchen. Better exhaust in the toilet. It is in favourably placed for better aeration and lighting.

Tobacco Smoke

Cigarette smoking is injurious to others health too. Typically about 500 ppm carbon monoxide is inhaled in each smoke which inactivate about 5% of hemoglobin. Other pollutants o cigarette smoke includes carcinogens such as aldehydes, tar, hydrogen cyanide, nicotine, lead etc.

The effects of cigarette smoking include indiscriminate lung cancer smokers and also others living in the environment.

Smokers recover slowly from acute virus infections. Cigarette significantly causes air pollution.

Definition Causes, Effects and Control Measures of Water Pollution

The physical and chemical constituent which alters the composition and characteristics of water is termed as water pollution.

Toxic chemicals, metals, suspended particles, radioactive products, extreme changes in temperatures, etc are considered as water pollution.

Sl.No.	Causes	Consequences	Control
1	Industrial effluents	Not portable and unagreable odour	Effluent treatment plants and treated well before discharge.
2	Domestic sewage	Life of aquatic life will be in danger due to depletion of oxygen	Town planning and better domestic water treatment plants
3	Thermal discharges	Bioamplification of toxic compounds.	Proper testing of soil and use of required quantity fertilizers.
4	Radioactive material dumping	Eutrophication, depletion of resoruces	Use of natural insecticides, predators to be encouraged.
5	Agricultural waste pesticides, fertili etc.	Spread of diseases.	Thermal discharges to be prevented.
6	Oil, grease and other washings.	Metal toxicity and health problem to women and children.	Anti-dumping laws to be made strict and followed.

Useful Constituents in Sewage

More than 99% of sewage is mere water that is polluted the 1% impurities with all the evil attributes. Hence water is the most abundant component of the wastewater to be put to reuse. The other components of wastewater and their potentials for recovery and reuse are as follows:

1. Organic matter is abundant in municipal sewage and some industrial wastes, particularly distilleries, sugar factories, food processing industries, slaughter houses, dairies etc. From this organic matter, manure and fuel gas can be realized. Anaerobic digesters constructed in distilleries may recover the entire cost in two, three years due to generation bio – gas that can replace the huge conception of coal. Also, as bio – gas is a clean fuel, pollution problem from the industry can be minimized.

2. Chemicals in industrial wastes can be varied depending on the industry. Some chemicals will have recovery value and can partly pay for the waste treatment processes. However the emphasis should be to reduce the concentrations of the effluents by suitable process alternations in the industry.

Methods of Recovery

Organic matter in sewage can be recovered by aerobic and anaerobic bacterial fermentation.

Setting tanks, activated sludge process, anaerobic sludge digesters, biogas plants etc. can enable the recovery of manure and fuel gas from organic components of sewages. The chemical in sewages are recovered by precipitation, adsorption, volatilization, chelation, solvent extraction, ion filters, reverse osmosis, etc.

Treatment of Sewage Water

Preliminary treatment

It is the process of removing suspended particles, coarse materials etc. by passing through bars or mesh screens.

Primary Treatment

In this step the fine particles, suspended impurities are removed by adding coagulants like alum $(Al_2(SO_4)_3)$.

Secondary or Biological Treatment

Biodegradable impurities are removed by aerobic bacterial digestion in specialized apparatus called Trickling filters or Activated sludge process.

Tertiary Treatment

Flocculation tank is used to remove phosphates. Ammonium salts are removed as ammonia by maintaining pH 11. Passed through charcoal to remove organic impurities and colour.

Sludge Recovery and Treatment

The sold bi-product formed due to the process of water treatment is called sludge. The useful constituents can be recovered and reused. Vermicompost method can be used for converting sludge into manure.

Reuse of Treated Wastewater

1. Where water scarcity is acute, the treated wastewater may be given further treatment (often called tertiary treatment) to make it potable. Even today many places treated sewage water is reused for domestic purposes including drinking.

2. Cooling towers in industries for heat exchange can use treated sewage water.

3. Irrigation and development of greenery can be largely benefited by treated sewage water.

4. Treated water can be used to recharge the ground waters. Soil systems have an excellent purification potential with rational management.

5. Pisciculture is possible in ponds formed by treated sewagewater, with careful management.

6. Sewage can be treated to any required quality and can be used for secondary purposes for industries. To cite an example, M.R.L. (Madras Refineries Limited) treats the municipal sewage of Madras by several advanced treatment processes to reuse it as the process water.

Definition Causes, Effects and Control Measures of Soil Pollution

Our Globe consists of 29 % land remaining eater. In which cultivable and habitable land surface is very less. Soil pollution is the result of urban technological revolution and speedy exploitation of every bit of natural resources. Making soil which can not be used for its normal sustainable use is soil pollution.

Sl.No.	Causes	Consequences	Control
1	Rapid Industrialization	Desertification and loss of agricultural productivity.	Industrial waste and bi-product dumping shall be prevented.
2	Population explosion and the activities	Spread of diseases, ground water contamination.	Solid waste treatment methods like vermiculture can be used
3	Modern agriculture	Bad odor, Spread of pathogens.	Useful constituents can be recovered and reused.
4	Dumpling solid waste	Food poisoning based on the agricultural products contamination.	Scientific disposal methods like burying in remote places can be practiced.
5	Mining and incidental activities.	Ecological imbalance, Many animals and plants can not survive.	Minimal waste producing technologies has to be invented.
6	Volcano, water logging, etc.	Pesticides, herbicides, cause nervous disorders, mutation etc in man.	Domestic solid wastes has to be segregated, categorized and used accordingly.

Definition Causes Effects and Control Measures of Marine Pollution

The process which makes the marine water and environment harmful to living forms and alters its ecosystem is termed as marine pollution.

In 1965, Lake Erie (USA) more than 80 tons of phosphate added daily. Each 400 g phosphate encourages about 350 tons of algal slime.

This growth appeared as big mounds, clogging pipes and interfered fishing and navigation.

- Heavy metals, chlorinated hydrocarbons, DDT, etc get accumulated in marine food and its products and finally reached to man and animals by food chain (Bioamplification).

- The oil spread due to oil spills forms a surface layer over water for several hundred square kilometre, which prevent the oxygen dissolution from the environment and hence marine living being die due to lack of oxygen.

- Radioactive nuclei having half life period with several million years are get in to mineral deposits or other living organisms. As the radiations are harmful which leads to may type of cancers in man and other animals.

- Heat causes local ecological imbalance in energy flow and biological equilibrium. In tropical countries animals live little lower than the lethal limit and increase in temperature leads to distress of even death.

Sl.No.	Causes	Consequences	Control
1	Dumping of metals like, mercury, lead, fluoride, etc.	Bio amplification and nutrient scarcity.	Discharges into sea should be controlled by stringent laws.
2	Dumping radioactive isotopes, and plastic wastes.	Death or decrease of marine life forms.	Dumping of radioactive and sold waste to be stopped.
3	Discharges containing dyes, industrial wastes, foaming agents, detergents, etc.	Eutrophication and decrease in dissolved oxygen and nutrient concentration.	Excessive use of fertilizers and washed into sea to be avoided.
4	Oil spill due to accidents, war, etc.	Coral reef population affected very much.	Thermal discharges to the sea to be prevented.
5	Thermal emission coming out from industries and thermal power plants.	Oil layer stay for longer period thus oxygen concentration of water drastically decreases.	Oil tanker collision to be avoided and better navigational guide to be adopted.
6	Community activities associated with recreation, tourism, etc.	Heavy metal poisons affect human consumers of marine products.	All type of Warfare in sea has to discouraged by international Laws.
7	Discharges carried by river, and other incoming water bodies.	Primary productivity decreases due to altered food chain.	Better check on pirates and their damage to sea vessels to be controlled.

Definition, Causes and Control of Noise Pollution

Unwanted sound is called noise. In USA noise is ranked second only to crime. In Japan more complaints are due to noise than any other. In India, sentimentally noise is considered to be an occasion. The presence of noise takes away the essence of music and speech. A rhythmic oscillation of frequency is not noise and is rather considered as music and enjoyable.

The research study of Indian Medical Research Council (1977-1982) revealed that about 10% of people in urban areas and 7% people in rural area have been affected complete of partial deaf due to noise pollution.

Noise is measured in decibel (dB). The zero on a decibel scale is at the threshold of hearing, the lowest sound pressure that can be heard. On this scale, 20 dB is a whisper, 40 dB is a quite office, 60 dB is a normal conversation, 80 dB for a bus, 100 dB for a train and 140 dB for a loud thunder.

The audible effect is in the frequency (Hz-Hertz, number of vibrations per second) range 20 Hz to 20,000 Hz. The sounds of frequencies less than 20 Hz are called infrasonic and greater than 20,000 Hz are called ultrasonic.

Sl.No.	Causes	Consequences	Control
1	Transport and public utility	Stress, aggressive, deafness and head-ache.	Reduction of source noise by better design and technology.
2	Industrial and service of equipments	Muscle pain, eye defect, nervous tension, heart attack	Noise absorbers like ceiling, screen, etc.
3	Thunder and mining explosion	Loss of attention leading to error, accident and thus less productive	Planting trees around noise and the concept is called green belt.
4	Construction and demolition site	Develop cracks in building and structures and lose its strength.	Use of ear plugs and other protecting aids.
5	Cultural and community activities.	Disturbs sleep and peaceful life.	Town planning and management.
6	Festival and religious activities.	Animals avoids mating and decrease reproduction.	Proclaim residential, hospital, school, court, as silent zone, and noise not exceed 40 dB.

Sources of Noise

Indoor Noise: It involves baby crying, playing of radios, banging doors, conversation, movement of furniture, kitchen noise, type writers, family fight, etc.

Outdoor Noises: Largest source is automobile traffic on the road. Others include railway traffic, loud speakers, nearby factory activity, etc.

Case Study

M.C. Metha vs Union of India and others (Yanni Musical Concert Case, 1988).

Yanni, great musician of America wanted to conduct a musical concert near Taj Mahal. The petitioner filled a writ under article 32 of Indian Constitution, preying for,

(i) Take action against authorities who destroying green belt around Taj Maha.

(ii) Direct State of U.P to shift the venue beyond 500 meters from Taj Mahal.

(iii) Not allow vehicles, generators and other sound producing equipments within 500 meters of Taj Mahal.

Judgement: Supreme Court fixed a distance of 750 meters from Taj Mahal and also to stay away the buses to that distance. It allowed battery operated vehicles upto 200 meters from Taj Mahal. An expert committee constituted to monitor the developments. As per section 3 of the "Environment (Protection) Rules, 1986, ambient air quality stands in respect to noise during the concert in the silence zone should not exceed 40 dB.

Thermal Pollution

It is physical pollution. Thermal pollution may be due to high temperature or low temperature discharges in water.

Sl.No.	Causes	Effects	Control
1	Atomic, and thermal power plants discharge	Reduction of dissolved oxygen.	Maximum utilization of thermal energy present in exhaust and industrial water.
2	Metallurgical industries discharges	Failure to of trout eggs to hatch and salmon to spawn.	Cooling ponds and towers to be installed and put in working condition.
3	At high temperature to the environment.	Mortality of fish due to reparatory or nervous failure.	Recycling of heat energy where ever possible for purposes like heating the room etc.
4	Volcanic eruptions, atomic explosions (nuclear test), war, etc.	Changes in physio-chemical properties of water.	Maintaining optimal temperature of process..
5	Nuclear accidents, Nuclear war, etc.	Increased biological activity of life forms.	Avoid over heating and wastage of heat.

Radioactive Pollution

Natural radiations like UV, cosmic rays can cause radioactive pollution. Artificial radioactive pollution is due to nuclear fusion and fission experiments. Radioactive materials degenerate naturally to non radioactive materials. It is an invisible danger.

Sl.No.	Causes	Consequences	Control
1	Radioactive mineral mining	Carcinogenic, mutagenic and tetragenic effects.	Controlled use of radioactive materials.
2	Nuclear fusion and fission experiments	Radioactive isotopes are incorporated in biological molecules.	Radiation therapy must be minimal and when absolutely required.
3	Nuclear accidents	Cancer and other related disease.	Safety norm in nuclear power plants must be followed strictly
4	Nuclear war	Prematured birth, birth defects.	Nuclear war must be avoided.
5	Radioactive therapy and medicine.	Damage of blood vessels, convulsions, vomiting, skin rashes, etc.	Terrorist and unauthorized agents acquiring radioactive material must be prevented.
6	Terrorist and unauthorized activities.	Plants losses its productivity and mutated	Open sea dumping of radioactive material should be stopped.
7	Nuclear waste dumping		

Radioactive Waste Management

Since there is no method of treatment awaiting natural decay is the only available alternative in handling radioactive wastes. The management aspects include: (1) Dilute and disperse, (2) Delay and decay and (3) Concentrate and contain.

If the quantity of waste is small it can be diluted and allowed to spread in the environment, so that the ambient concentrations of radioactivity are within the permissible range. Where the half life period of the radioactive waste is low, the waste can be stored with proper shielding till the material decays to a safe level, after which the waste can be disposed. For wastes with large half life periods, the volume is reduced by concentrating the pollutant and the material is encased in a radiation shielding container like lead containers or concrete casing blocks which are to be buried at a suitable place. All the three approaches are bound to increase the back ground levels of radioactivity and also pose a hazard in case of a disaster like and earthquake or even sabotage. During recent times radioactive materials are being pilfered by terrorists. This is also a serious threat for the well being of mankind.

Solid Waste Management: causes, effects and control measures of Municipal solid wastes

Municipal solid wastes are material which is discarded by human beings in public or protected places. It is generated due to human activity. It includes kitchen wastes, broken items, plastic and leather products, medical wastes, agricultural bi-products, human excreta, etc.

Causes of Municipal Solid Waste Generation

- Human normal life activity
- Luxurious activity like washing machines, televisions, computers and discarding after its use.
- Special activities of children, diapers, breaking items, toys, game materials, etc.
- Medical containers, injection needles, etc.
- Plastic and other forms of carry bags.
- Food wastes and vegetable peels, etc.

Effect of Municipal Solid Wastes

Municipal solid waste management is the part of urban planning. It is because if not it causes serious environmental damage.

Dumping on land surface leading to wastages of precious land.

- Health hazards like spread of diseases.
- Bad odor and related health problems like head-ache.
- Ground water contamination since the waste is carried inside during rainy season.
- Environmental pollution due to fermentation and emission to the environment.

Control Measures of Municipal Solid Wastes

The methods used to manage sold wastes are

a. Waste reduction
b. Collection and transfer
c. Composting
d. Incineration
e. Landfills
f. Special wastes.

a) **Waste Reduction:** It is basically awareness among human population. It is asked to reduce the solid waste generation as much as possible. The common method used is 3R policy.

3R Policy: 3R Stands for Reduce, Reuse and Recycle. It means as much as possible reduce the amount of solid waste formation. It involves maximum utilization of vegetables and other products. When ever possible reuse the items. It includes reusing plastic carry bags, covers, etc. Further water bottles, plastic cans, broken toys, milk bags, containers, etc., instead of throwing them in open environment store them safely and give it to rag pickers. It constitutes recycle process.

Thus reduction at source drastically minimizes the cost of waste treatment in addition sustainable use of precious raw materials and finished products. In addition it saves the money and economic income to the family without additional investment.

b) **Collection and Transfer:** It is the process of accumulating all the waste and transport to place of processing and disposing safely. Indeed collection has to be done in a rigorous and careful manner. For this purpose inhabitants are educated to know the biodegradable and non-biodegradable solid wastes.

Biodegradable and Non-biodegradable Solid Wastes: Biodegradable solid wastes are end wastes which can be subjected to biological process like fermentation. Wastes like wasted food, vegetable peels, etc.

Non-biodegradable wastes are potent environmental pollutants. They can not be degraded easily by biological treatment. It includes glass materials, plastic carry bags, toys, metallic products, leather items, tires, tubes, etc.

Indeed separate bags with different colour codes are designated for both the category of wastes. They are collected separately. The process is called as source separation.

Medical wastes are segregated and collected separately from hospitals. This is because of potential microbial contamination. Recent times it is advised to have own solid waste management system for every hospital.

Techniques of Collection of Solid Wastes

(i) **"Just-in-time" collection:** Some cities use "just-in-time" collection systems, where residents bring out their wastes at the time the collection vehicle reaches a certain spot and signals its presence. This system reduces the health hazards associated with wastes on streets and roadsides, and prevents unauthorized waste picking.

(ii) **Communal collection**: Communal collection, which is very common in developing countries, involves individuals bringing their waste directly to the collection point, usually a container that can be accessed by foot.

An advantage of communal collection points for drop-off of household waste is that these facilities provide more or less continuous access to disposal or materials recovery facilities. Disadvantages of communal collection arise from the fact that such facilities may receive little attention from municipal authorities. Additionally, residents may deposit dangerous materials in or near the container.

Ideally, communal containers should be designed to prevent animals from getting access to the contents. While tall containers or those with small openings or heavy lids may accomplish this purpose, they are also difficult to use, especially for children.

(iii) **Special collections:** Special materials, such as bulky items, white and brown goods (old appliances and electronics), furniture, leaves, construction materials and tree stumps, must often be collected separately, due to their size and the fact that they are generated irregularly.

Overview of Transfer

Transfer refers to the movement of waste or materials from the primary collection vehicle to a secondary, generally larger and more efficient, transport vehicle. Now a days collection is carried out from each house and them loaded into a larger containers which is transported by closed tanks.

c) Composting Solid Waste

Centralized composting at the municipal scale: It refers to composting of animal and plant wastes from multiple sources, where the wastes are transported from several points to a facility that can receive 10 to 200 tons per day.

Volume reduction in composting: The action of the bacteria transforms materials into gases. Insects and microorganisms are allowed to feed on the organics. Additional volume reduction occurs due to removal of non-compostables during pre-processing or final screening, in addition to the moisture loss and volume reduction during composting itself. Thus a 100-ton-per-day facility will be reduced to only 30-50 tons of compost per day.

Small-scale composting of animal wastes: Composting and digestion of bones is carried out as a small industry in some developing countries. It can produce ingredients in the manufacture of fertilizer, animal feed, and glues. The traditional methods of sun-drying, breaking up bones manually, composting in pits (sometimes with the addition of household organics), and steam digestion carry various health risks, and cannot be considered a sound practice.

Small-scale aerobic composting of animal wastes, including manures, hide scrapings, and tannery and slaughterhouse wastes can also produce fertilizers.

Vermiculture for Waste Recycling: Currently the most popular and widely employed technique for solid waste disposal relies on earthworms generally referred to as "farmers' friend". For centuries, earthworms, as biological natural agents, have been in the business of decomposing wastes and enriching the soil structure.

Typically earthworm is allowed grow in the solid waste. It takes the waste and converted into orgamnic manure when it pass through its digestive system. Present days bioreactors for cost-effective and environmentally sound waste management. The product generated is organic fertilizer and used as a manure for farming.

Anaerobic digestion: Anaerobic digestion is an approach to composting, which shows promise as a sound practice in industrialized countries. A number of facilities in France and Belgium appear to be capable of composting mixed waste under pressure and are recovering both compost and methane gas.

Wastewater sludge and human fecal matter: They are high-nitrogen materials that can be aerobically composted under certain circumstances. They are high in moisture, sometimes actually liquid. Composting can only work if they are combined with carbon sources such as wood or paper and bulking agents, such as chipped wood or rubber, which are dry and maintain air spaces in the compost.

Anaerobic digestion of these materials can also work, and is a sound practice on farms in industrialized countries. It is operationally more difficult than aerobic composting. It produces sledge as dry into cakes, which may be used as fertilizer.

Manures and animal wastes: For centuries animal wastes are used as manures. This is due to high in nitrogen content. The bedding materials like straw, wood, plant wastes, or paper are easy to compost to manures. Recent times vermicomposting (worm culture) is used to make organic manure. It is excellent for sustainable agriculture.

Incineration of Solid Waste

The primary benefit of MSW incineration is a substantial reduction of the weight (up to 75%) and volume (up to 90%) of solid waste, which can be valuable if landfill space is scarce. It is basically burning in controlled condition and harvest the energy out of it.

Energy production: In waste-to-energy plants, heat from the burning waste is absorbed by water in the wall of the furnace. It is allowed to act on turbines to generate electricity.mAbout one-fifth energy produced is used at the facility for general operations. The remaining electricity is sold to public and private utilities or nearby industries.

Steam production and use: The energy generated by European waste-to-energy plants typically goes to supply steam to district heating loops. Steam generated in incineration facilities can also be used directly by a customer for manufacturing operations. Steam generated in an incinerator is supplied to a customer through a steam line, and condensed steam is sometimes returned by a separate line.

Cogeneration: Combined production of steam and electricity is referred to as cogeneration and can occur in two ways. If the energy customer requires steam conditions (pressure and temperature) that are less than the incineration plants design specifications. A turbine-generator is used to produce electricity and thus reduce steam conditions to appropriate levels for the customer. Or, if the steam purchaser cannot accept all the steam produced by the facility, the excess can be converted to electricity.

d) Landfills

It is also known as dumping. The solid waste which is generated after incarnation or the one can not be used for any of the useful purpose is dumped under the soil. Usually landfill area is remote area designated for the purpose. In general waste is contained under the land within concrete structure in order to avoid water contamination and other problems in the surrounding area.

Types of Special Solid Wastes and Management

Medical waste

Sound practices for managing medical wastes are characterized by:

Source separation within the hospital: Isolates infectious and hazardous wastes from non-infectious and non-hazardous ones, through color coding of bags or containers.

Take-back systems: Where vendors or manufacturers take back unused or out-of-date medications for controlled disposal.

Tight inventory control over medications: To avoid wastage due to expiration dates (really a form of waste reduction).

Proper disposal of hospital wastes: Final disposal should preferably be done in a specially designated cell, which should be covered with a layer of lime and at least 50 cm of soil. When no other alternative is available for final disposal, hospital wastes may be disposed of jointly with regular wastes. In this case, however, hospital wastes should be covered immediately by a meter thickness of ordinary MSW and always be placed more than two meters from the edge of the deposited waste.

Household Hazardous Waste

Households generate small quantities of hazardous wastes such as oil-based paints, paint thinners, wood preservatives, pesticides, household cleaners, used motor oil, antifreeze, and batteries. Household hazardous waste in industrialized countries such as the US accounts for a total of 0.5% of all waste generated at home. In developing countries the percentage is even lower.

There are no specific, cost-effective sound practices that can be recommended for household hazardous waste management in developing countries.

Household Hazardous Wastes Management

- The priority wastes are identified with reference to the damage they do when released into the environment. For example, the prioritisation of battery collection would be related to a high reliance on incineration, for which heavy metals are particularly unsuited.

- There is ample, clear, and frequent public education as to the need for separation and the available opportunities to separate.
- The separate disposal or recycling opportunity is convenient, matched to the normal disposal opportunity, and either continuous or frequent.

Tires

Reuse through retreading for extended service; shredding and grinding for use in road paving material; and cutting them up for use as padding in playgrounds and buffers on railway tracks.

Thermal destruction in cement kilns with consequent energy recovery. This system requires adapting cement kilns to receive solid fuels.

Processing in pyrolytic ovens. As a result, the process can be relatively expensive and will become cost-effective only when the accumulation of tires becomes hazardous due to potential fires or expensive due to conflicting land use.

Used Oils

Used oils are generated mostly in gas stations and in mechanics' shops. These oils frequently enter the sewage system, causing problems in treatment plants or in the receiving water bodies. When oil is collected haphazardly as part of the MSW stream, it causes problems at the landfill and often becomes part of the landfill leachate.

Re-refining into lubricating oil. Afgter some primary treatment can be used for lubrication either directly or mixed with fresh oil.

Use as a fuel. The used oil is best used as a fuel in cement kilns where heavy metals are absorbed into the cement matrix.

Wet Batteries

Recycling in controlled environments. It is important to note that small-scale recycling is typically highly polluting and should be avoided.

Drainage of acid with subsequent neutralization, and melting of the metal casing in a nonferrous foundry.

Construction and Demolition Debris

Generated regularly in urban areas as a result of new construction, demolition of old structures, and regular maintenance of buildings. These wastes contain cement, bricks, asphalt, wood, and other construction materials which are typically inert. In addition lot of noise also generated.

Inventory control and return allowances for construction material. This ensures that unused materials will not get disposed of unnecessarily.

Selective demolition. This involves dismantling selected parts of buildings to be renewed.

On-site Separation Systems: Segregate the materials like bricks, stones, gravel, iron rods, etc and use or dispose accordingly.

Crushing, milling, and reuse of secondary stone and concrete materials. Major part can be used to approved road construction materials specifications. The remaining items have to be disposed according to the specifications.

Industrial Waste

Waste generated in industrial settings has non-hazardous and hazardous components, with non-hazardous waste representing the greater part of the volume. The hazardous component of this waste, while being relatively small in volume, can pose very large environmental and health problems.

Sound practices are extremely varied for industrial waste. In any case, all sound practices must lead to separating hazardous industrial waste. In those cases where municipal authorities are forced to provide a temporary solution for the disposal of such waste, specially designed cells should be provided within the municipal landfill. These cells must be isolated so that the waste pickers cannot come into contact with industrial waste.

Ecosanitation and Solid Waste Management

The Foundation with its involvement in biotechnology has successfully developed solid waste management technologies for converting household and segregated city waste to organic manure through a process of sanitization, deodorization and accelerated decomposition.

The Ministry of Urban Development and Poverty Alleviation (CPHEEO) recently short listed Morarka Foundation to provide know how for conversion of city waste into vermicompost under Supreme Court direction for all municipal towns having more than 10,00,000 population as an "Appropriate Technology On Solid Waste Management".

Role of Individual in Prevention of Pollution

Buy and use ecofriendly products. Earthen pot is the ecomark for ecofriendly products.

Energy Saving

- Use energy efficient methods in day today activities like heaters, vehicles with greater mileage etc.
- Put off power for light, fan, etc when it is not absolute required or not in place of it s use.
- Use energy efficient fluorescent lambs.
- Construct houses and other using places where maximum utilization of day light and atmospheric air.
- Avoid using air conditioners rather design controlled airy rooms.

Pollution Control

- Not to pollute his own surrounding and also neighbours environment.
- Use private vehicles when it is absolutely necessary, rather prefer community transport.
- On road put off the engine of vehicle while in traffic signal, traffic jam, etc.
- Keep the vehicle fit and frequent pollution monitoring.
- Use better quality oil and fuel, do not go for adulterated cheap ones.
- Do not dump solid and liquid house hold sewage openly.
- Segregate waste before disposal and dispose appropriately.
- Obey municipality regulations and law of the land.
- Avoid the use of open toilet.
- Eat limited and required amount.
- Try to implement to plant tree based on the motto "One Man One Tree".
- Reuse and recycle materials especially materials especially plastic products in addition to their optimum used.

Water Utilization

- Do have any activity, which pollute community water resources like pond, river, etc.
- Minimize the use of water and where ever possible recycle.
- Close the water taps after use of toilets, bathrooms etc.
- Keep the check of waterways and taps for nonleakage.
- Have rainwater harvesting system.
- Use good quality water when it is required.

Pollution Case Studies

Love Canal (Soil Pollution)

Love Canal is a 16 acre land fill in a residential neighbourhood of Niagra Falls, New York, USA. It was intended to be a canal running at the center of the Inductrial Empire but was never completed and was finally used as a dump-yard. In 1942, the city of Niagra Falls and Hooker Chemical and Plastics Corporation (HCPC), a local industry started dumping wastes in to the canal. By 1950, about 25,000 tons of chemical wastes including highly toxic pesticides, herbicides and other chemicals were dumped in the abandoned canal. In 1950s, as the city of Niagra Falls grew rapidly, the neighbours complained of foul odours and rats.

A school was built right on the top of the chemical. Children always enjoyed by playing with balls wetted by 'love canal'. After nearly 20 years of use, people complained of children with birth defects of chronic medical problems like asthma, bronchitis, infections and hyperactivity. Dogs also suffered from loss of hair and skin diseases.

Disputes public rallies, negotiations and law-suits continued for six years and finally HCPC agreed to pay 250 million dollars as compensation to 'Love Canal' residents.

Minamita Disease (Soil Pollution)

In the early 1950s people of Minamita, a small coastal town of Japan notices a strange behaviour that they called 'dancing cats'. Their cats were twitching, stumbling and jerking about as if they were drunk, as they were suffering from brain damage what is now known as 'methyl mercury poisoning'.

The Chisso Chemical plant used to release residues containing Mercury into The Minamita Bay. As the element mercury is not soluble in water, it was assumed that it would sink into the bottom sediments and remain inert. However, bacteria living in the sediments were able to convert the elemental mercury into soluble methyl mercury which finally absorbed by the tissues of aquatic organisms. Fish containing about 50 ppm of mercury has been reported and the allowable limit is 0.05 ppm.

In humans epidemic of nervous problems including numbness, tingling sensations, headaches, blurred vision, slurred speech and loos of muscle control. Abnormally high birth defects such as paralysis, mental retardations were also reported. After 20 years of protests, the Chisso Company agreed to pay compensation.

Chernobyl Nuclear Disaster (Nuclear Pollution)

On April 26, 1986, the nuclear reactor of Chernobyl power plant (USSR) exploded due to predominantly human error. The explosion occurred due to high pressure developed inside the reactor and hence the 1000 tons steel blew off. The causality and radiation exposure are under estimated. The fact is the radiation spread over many neighbouring countries like Denmark, Suden, Norway, etc. Children and adults are most affected people. The agricultural products exposed to radiation developed cancer, thyroid abnormalities in human.

26 January 2001 Bhuj Earthquake, Gujarat, India

The Bhuj earthquake that shook the Indian Province of Gujarat on the morning of January 26, 2001 (Republic Day) is one of the two most deadly earthquakes to strike India in its recorded history. One month after the earthquake official Government of India figures place the death toll at 19,727 and the number of injured at 166,000. Indications are that 600,000 people were left homeless, with 348,000 houses destroyed and an additional 844,000 damaged. The Indian State Department estimates that the earthquake affected, directly or indirectly, 15.9 million people out of a total population of 37.8 million. More than 20,000 cattle are reported killed. Government estimates place direct economic losses at $1.3 billion. Other estimates indicate losses may be as high as $5 billion.

Many non-governmental organizations are involved in relief and rehabilitation efforts. International organizations have also rushed relief materials and medical personnel to Gujarat.

Bhopal Tragedy (Air Pollution)

Around 1 a.m. on Monday, the 3rd of December, 1984, in a densely populated region in the city of Bhopal, Central India, a poisonous vapor burst from the tall stacks of the Union Carbide pesticide plant. This vapor was a highly toxic cloud of methyl isocyanate. Of the 800,000 people living in Bhopal at the time, 2,000 died immediately, and as many as 300,000 were injured. In addition, about 7,000 animals were injured, of which about one thousand were killed. "A series of studies made five years later showed that many of the survivors were still suffering from one or several of the following ailments: partial or complete blindness, gastrointestinal disorders, impaired immune systems, post traumatic stress disorders, and menstrual problems in women. A rise in spontaneous abortions, stillbirths, and offspring with genetic defects was also noted."

The Indian Government, in response to the tragedy and pressure from the Indian people, filed a compensation lawsuit against the UCC for an estimated $3 billion. In October of 1991, the Indian Supreme Court upheld a settlement, which had been appealed from a lower court decision of 1989, under which Union Carbide had to pay $470 million in compensation of all claims. However, Union Carbide declared that the company had no intention of doing anything further for the victims.

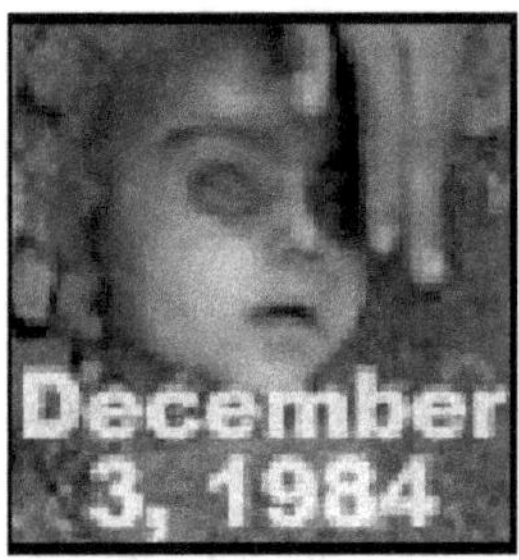

Oil Spillage

Contamination of coastal amenity areas is a common feature of many spills leading to public disquiet and interference with recreational activities such as bathing, boating, angling and diving.

Industries that rely on a clean supply of seawater for their normal operations can be adversely affected by oil spills.

Oil spills can have a serious economic impact on coastal activities. Oil spills may kill or injure hundreds or even thousands of marine mammals and birds. Further, it can depress the immune system of marine mammals, making the animals susceptible to diseases they could normally fight off. Polar bears in Svalbard,

Blowout of exploratory well Ixtox 1 in 1979. When workers were able to stop this blowout in 1980 an estimated 140 million gallons of oil had spilled into the ocean.

Oil tanker Mega-Borg released 5.1 million gallons of oil as a result of an oil transfer accident.

According to Ocean Planet there are 706 million gallons of oil pollution in a given year.

Disaster Management

Disaster is an unexpected event which has direct impact on human and other life forms immediately. Disaster management is also known as mitigation process.

Flood Management

- Be aware and listen to communication links.
- Stay at higher area of the house if trapped inside house.
- Wear simple dress and alert to the situation.
- Help the elder and younger one's in the family.
- If sufficient after flood warning, run to upside of the living area.
- Keep informed to service persons and NGO who are around to help the people.
- Not to stay near the weak structures or trees since they may fall on you.
- Keep battery operated radio with you. Also other communication devices like cellphone.

Earthquake Management

- Stay Calm!
- Follow the Earthquake Family Plan.
- Duck and Cover!
- Do not try to enter or leave any buildings.
- Secure all objects capable from falling during an earthquake.
- Put all breakables in fastened cabinets.
- Buy flashlights, batteries, and battery powered radios and televisions.
- Have several gallons of bottled water on hand. Have canned foods available.

Landslides Management

If you are living in a vulnerable area consult a professional geotechnical expert for opinions and advice on landslide problems and on corrective measures you can take.

Learn to Recognize the Landslide Warning Signs

- Doors or windows stick or jam for the first time.
- New cracks appear in plaster, tile, brick, or foundations.
- Outside walls, walks, or stairs begin pulling away from the building.

- Slowly developing, widening cracks appear on the ground or on paved areas such as streets or driveways.

Make Evacuation Plans

Plan at least two evacuation routes since roads may become blocked or closed.

In case family members are separated from one another during a landslide or mudflow this is (a real possibility during the day when adults are at work and children are at school), have a plan for getting back together.

Stay Away from the Slide Area

There may be danger of additional slides. Check for injured and trapped persons near the slide area. Give first aid if trained. Remember to help, who may require special assistance to infants, elderly people, and people with disabilities. Listen to a battery-operated radio or television for the latest emergency information in case of major slides.

Cyclone

A violent storm, rotating about a calm center of low atmospheric pressure. This center moves onward, often with a velocity of twenty or thirty miles an hour.

Preparations: Before the onset on cyclone season, stock up on such items as candles and matches, cooking gas, non-perishable foods, and bottled water. Make sure that your flashlights and radio have fresh batteries. When a Class I cyclone is announced, you will want to make sure that your car's gas tank is full. If a Class II warning is called, start boiling water and make sure that you have sufficient supplies to last you in case power is out for several days (food, water, candles, etc. Once a Class III warning is issued you should stay off the roads.

Information: The best way to stay informed is to listen to the Radio and Television if it is still working.

Emergency Procedures: In the aftermath of a serious cyclone, communication and transportation networks are likely to be severely damaged. Froemergency assistance contact with the relevant authorities or get in touch with the nearest police station.

Tsunami

Tsunami is a Japanese word with the English translation, "harbor wave." In the past, tsunamis were sometimes referred to as "tidal waves" by the general public, and as "seismic sea waves" by the scientific community. The term "tidal wave" is a misnomer.

Cuase: Earthquake under the sea is reason for Tsunami. Tsunamis are unlike wind-generated waves, can have a wavelength in excess of 100 km and period on the order of one hour. It travels at about 200 m/s, or over 700 km/hr. They can also travel great, transoceanic distances with limited energy losses. The earthquake-generated 1960 Chilean tsunami, for instance, travelled across over 17,000 km across the Pacific to hit Japan.

The Tsunami of 26 December 2004 in the Bay of Bengal and the Indian Ocean

The tidal waves occurred because the massive earthquake heaved the ocean floors off the Indonesian coast. "The undulation of the ocean floor triggered the tsunami," says Shetye. "Once that happens, a tsunami propagates at the speed of 750-800 km/hr. The waves would have travelled at the speed of a jet engine to hit Sri Lanka and India within just over two hours after the quake." From the sky or coastline, the open sea would still appear unsuspectingly calm. As the waves approached the southern and eastern coastline of India, they slowed down but rose in height. In the shallow coastal waters the waves morphed into a wall of water almost 10 m high, swallowing everything in its way.

The official government toll in south India from tsunamis rose slightly late on Friday to reach at least 10,001 people confirmed dead and 5,689 missing, most presumed dead.

Tamil Nadu suffered the heaviest casualties with 7,941 dead while neighbouring Pondicherry lost 583 people. The ministry said 170 people died in Kerala and 105 in Andhra Pradesh.

In the worst-hit district of Nagapattinam, in Tamil Nadu, the total dead hit 6,035. The ministry says final confirmation is still awaited before it can list all the missing as presumed dead and the hunt will go on through next week.

Questions

1. What are the causes and its effects of pollution in the environment?
2. How will you control pollution in
    a) air
    b) Water
    c) Soil
    d) Sea
3. What is noise pollution? Explain its effects and control methods.
4. Explain the effect of thermal pollution for aquatic animals?
5. Explain nuclear pollution with a suitable case.
6. What is a solid waste and explain its effect on soil. How will you manage solid waste?
7. What are roles of a citizen in reducing pollution?
8. How will you manage flood, earthquake, cyclone and landslides

Unit III

Natural Resources

Forest resources: Use and over-exploitation, deforestation, case studies- timber extraction, mining, dams and their effects on forests and tribal people – Water resources: Use and over-utilization of surface and ground water, floods, drought, conflicts over water, dams-benefits and problems – Mineral resources: Use and exploitation, environmental effects of extracting and using mineral resources, case studies – Food resources: World food problems, changes caused by agriculture and overgrazing, effects of modern agriculture, fertilizer-pesticide problems, water logging, salinity, case studies – Energy resources: Growing energy needs, renewable and non renewable energy sources, use of alternate energy sources. case studies – Land resources: Land as a resource, land degradation, man induced landslides, soil erosion and desertification – role of an individual in conservation of natural resources – Equitable use of resources for sustainable lifestyles.

Environmental Resources and Natural Resources

Natural resources are naturally occurring one having direct economic use for human population. Eg. Land, water, sea, etc.

Environmental resources are components of the environment. It has intrinsic value of their own. Eg. Bio diversity of plant and animal populations, hotspots, etc.

Renewable and Non Renewable Resources

The resources that can be replenished with relatively ease of time are referred as renewable resources. Solar energy is a best renewable energy and it can last atleast 6.5 billion years. Others include fertile land, human beings, etc. Non renewable resources are available in a definite quantity and are exhausted after a certain period of time. They include minerals, fossil fuels, extinct life forms and endemic and endangered species.

Forest Resources

A forest is defined as an ecosystem dominated by trees and other woody vegetation having rich biodiversity.

Environmental Functions

Protective functions like prevention of soil erosion, droughts, floods, intensive radiation, noise, etc. Productive functions like wood, gum, fibre, honey, paper, etc. Accessory functions including recreation, aesthetic, educational, tourism, etc. Forest is the place where highest production of biomass takes place. In environmental oxygen-carbon dioxide balance forest plays a major role.

Ecological Significance of Forest

Forest prevents ecological degradation by maintaining through food chains and other biological cycles.

It plays vital role in maintaining global climate, hydrolytic cycle, etc. It maintains and support biodiversity. Forests preserve oxygen carbon dioxide balance in the environment.

It is estimated that a typical tree provides about $ 200,000 worth of benefits annually.

Non-extractive Uses of Forests

Sl.No	Use	Description
1	Gathering	This includes plant materials for food or other uses. Mushrooms, young spring greens, reeds (for baskets) are some important examples
2	Hunting and Trapping	This includes both large and small game, hunted or trapped for food or for skins. Game species like bear, wolf, fox, squirrel, and various bird species such as grouse, ptarmigan, ducks, etc.
3	Fishing	Fishing is classified as sport fishing or commercial. People practices fishing along the banks of forest.
4	Tourism	Forest tourism includes all of the categories listed under recreation, plus guided nature walks and boat tours. Tourism is a growing source of revenue
5	Recreation	Recreation can include bird watching, boating, hiking, skiing, and mashing
6	Medicine	In India medicinal plants were known from Vedic times. Their properties were described in Rigveda and Atharva Veda. On the basis of chemical constituents, alkaloidal are used as sedatives. For example, Senna and Rhubarb are purgatives, while Arjuna, Aswagandha (sedative) are cardiac drugs. Taxol is a drug extracted from plant *Taxus beccatum*. *Azadirachta indica* (Neem) leaves contain alkaloids which are used for Carminative, emetic, anthelminthic tooth paste preparation. *Ocinum sanctum* (Tulsi) leaves contain volatile acids useful for Cough, cold, gastric disorder treatment.
7	Minerals	Forests are rich sources of mineral deposits and fossil fuels
8	Oils	Forest products also include edible and luxury oils like sandal wood oil and Eucalyptus oil, etc. Bio-diesel is the oil obtained from Jetropha seeds.
9	Shelter	Forest is used as a traditional dwelling place by human population all over the world and in India. Tribal people still live and prefer use it for their habitation.

Economic Significance of Forest

Industrial wood – timber, round wood, match & pulpwood. Raw materials-paper industry, ores, etc. It provides firewood. Minor Forest Products–bamboo, fodder, lac, sandalwood, honey, resin, gum, tendu leaves, etc. are in regular use by population.

Income from forest resources is aggregated under the head of "Forestry & Logging" under income from agriculture.

Overexploitation of Forest

Overexploitation means extracting forest resources over it self-sustainability. Forest is rich in resources, which is destroyed by human activity due to overexploitation.

Overexploitation is purely human activity. This leads to degradation and destruction of the resource either temporarily or permanently.

Sl.No.	Causes	Effect	Control
1	Increased population and their survival needs	Soil erosion and flooding	Clear policy where forest can not be used for economic purpose.
2	Government encouragement for foreign currency and tourist economy.	Decrease in forest cover leading to the ecological imbalances like climate change.	People who depend on forest for life shall have be educated to minimal and optimal use.
3	Industrialization beyond the requirement and export to foreign countries.	Change in season pattern and rain pattern in addition to decrease in rain.	Decrease the use of forest derived raw materials for industrial purpose.
4	Human greed and excess needs	Deforestation, Desertification, etc.	Trade of forest products shall be banned.
5	Survival needs of layman and labourman who depends on forest.	Loss of biodiversity.	Reforestation policies has to be strengthened if any deforestation by any reason.

Deforestation

Deforestation is the loss of forest cover due to natural or man made causes. According to the World Resources Institute, 78 % of the world's original frontier forests have already been destroyed or degraded, much within the past three decades.

Reason for Deforestation

- Mining, urbanization, industrialization.
- Population growth and poverty which drives the people to occupy forest.
- Government role in giving title to landless tribal people.
- Over exploitation of natural forest resources.
- Overgrazing, fire wood extraction, Timber extraction, etc.
- Developments like laying roads, railway tracks etc. Agricultural purpose, forestland is converted into cultivation area by deforestation.
- Government also encourages deforestation for the purpose of economic balance in global scenario, like tea plantation, etc.
- Natural calamities like forest fire, volcanic eruptions, earth quake, cyclones, etc.

Effect of Deforestation

Global Warming

The equilibrium in the biological cycle is getting affected. For example carbon dioxide get accumulated in the environment due to the lack of availability of sufficient forest cover to handle the gas by photosynthesis.

Loss of Biodiversity

Over 70% Plants and animals living in around the forest Deforedtation leads to l ose their habitat, protection and livelihood. The consequence is species loss, imbalance in food chain and nutrient cycle.

Soil Erosion

The productive soil surface is washed away by water during rainy or windy season. This reduces the growth of plants in and around the forest. In addition agricultural productivity of land around forest area get decreased.

Deforestation and Tribal People

Tribal people live and obtain their livelihood from in and around forest. They perform by hunting, gathering, trapping and slash, and burn cultivation. The estimates show 5% of population live and enjoy forest.

- Tribal people representing about 5000 cultures are vanishing along with the intensity of deforestation.
- Deforestation leads to urbanization and tribal people continue to practice their 'right to practice' their culture, which get damaged in the urban environment.
- They are affected mentally and physically in the absence of forest cover.
- Many Tribal people lost their home and energy needs in addition to their native culture.

Chipko Movement

In India, local women decided to fight the government and the vested interests to save trees. The women of Chamoli District, Uttar Pradesh, declared that they would embrace-literally "to stick to" (chipkna in Hindi)-trees if a sporting goods manufacturer attempted to cut down ash trees in their district.

Since initial activism in 1973, the movement has spread and become an ecological movement leading to similar actions in other forest areas. The movement has slowed down the process of deforestation.

Chipko adapted the slogan 5F, includes Fodder, Fuel, Food, Fertilizer and Fibre.

Timber Extraction

Timber extraction is the process of felling trees from forest in particular for economic purpose. It is for construction of buildings, bridges, railway tracks, house building, furniture, sleeping cots, paper industry, extraction of medicine, etc.

Effect of Timber Extraction

- Large scale timber extraction leads to deforestation.
- It leads to loss of biodiversity.
- Urbanisation and due to migration of tribal people.
- Soil erosion and landslides.
- Ecological imbalance and global warming, climate change, etc.

Remedies for Timber Extraction

- Alternative sources for timber.
- Change of technology where timber is used as raw material.
- Minimal use and reuse and recycle where ever possible.
- Forest economy due to timber extraction to be discouraged.
- Paper industry to look for alternative raw material instead of timber.

Mining

Mining is an activity for the purpose of extraction, concentration, and smelting of economic minerals from a mineral deposit.

It includes exploration development constructing the mine and mining.

Types of Mining

(i) Traditional hard rock mining usually involves digging tunnels and adits (horizontal entrances into hillsides) to reach lodes of mineral-rich ore.

(ii) Traditional underground mining and open pit mining require that the ore-bearing rock be removed and then put through a milling and extraction plant to extract the desired minerals.

(iii) Surface mining is the extraction of minerals from the earth surface.

Effects of Mining

- Deforestation is the immediate consequence of mining in forest area.
- Displacement of people and rehabilitation problems.
- Mining process releases harmful toxins, like mercury and cyanide which are often used in the gold extraction process.
- The chemicals used for mining contaminate local water bodies and rivers.
- Nose pollution and disturbs the living forms in the surrounding area.
- Excessive human activity leads to loss of territory for the forest animals.
- They clear trees and detonate explosives on hillsides, which causes erosion, landslides, etc.
- Urbanisation due to set up industries near mining area. This leads to clashes with the local indigenous peoples over land ownership, result in fatalities.

Dams and Their Effects on Forest and Tribal People

Dam is a barrier built across flowing water in order to create a water reservoir. Major dams are built across large rivers to save and use natural water effectively. Building dam is an ancient origin when Kings ruled the country.

After independence India is in the urge of self-dependence of food and energy, which leads to large scale construction of dams. At present India is blessed with more than 1550 large dams many small water reservoirs. Tehri dam is considered to be the highest built across river Satluj, in State of Maharashtra.

Dams and Its Effects on Forest

- Large hydroelectric projects, funded by international aid and development organizations like the World Bank, have lead to widespread forest loss in recent years.

- It led to killing off local wildlife, destroying aquatic habitats and affecting fish populations, displacing indigenous peoples, and adding more carbon to the atmosphere as the submerged wood rots.
- The forest equilibrium and balance is lost since the part of the forest is lost. The other part of the forest is depending upon the lost area of the forest.

Effect of Dams on Tribal People

- Tribals are endangered species in India and the loss of such biodiversity cannot be tolerated.
- It is the Protection given in Indian Constitution to tribal people and hence their culture cannot be questioned and destroyed.
- Right to life of tribals in their biotic environment is a guaranteed one and the dams dampen this right.
- The displacement and cultural change affects the tribal people both mentally and physically. They are not adopted to the modern food habits and life styles.
- They are ill-treated by modern society beyond the protection of law and least considered in any forum.
- April 18 is considered to be World ethnic day.

Water Resources

Water is a major constituent on earth's surface. Water is an Inorganic compound containing two atoms of hydrogen linked with one atom of oxygen. The component of rain is water. Water makes the earth livable for living organisms. Hydrolytical cycle play a major role in climate maintenance. Water is a renewable resource.

With the exception of ammonia, water has the highest specific heat capacity (4184 J/kg/K). Its heat of vaporization is also high (2258 kJ/kg). Water is a polar protic solvent and acts as a solvent and medium for many processes.

Water is a key resource vital for life activities of all living forms. Water plays an important role in the manufacture of essential commodities, generation of power, transportation, recreation and in domestic, commercial, agricultural and industrial activities.

Importance of Water

- Development of earth's surface and moderating climatic conditions and diluting pollutants.
- All physiological activities of living forms are carried out in water environment.

- It is the component of photosynthesis in which solar energy is trapped for food.
- All industrial process requires water one time or other.

Quantity of Water

The total quantity of water available in the universe is 1360,000 x 10^{12} m³ out of which 97.3 % is in oceans.

About 2.7 per cent of the total water available on the earth is fresh water of which about 75.2 per cent lies frozen in polar regions and another 22.6 per cent is present as ground water. The rest is available in lakes, rivers, atmosphere, moisture, soil and vegetation.

Water Distribution

Uneven and irregular distribution of water is noticed among continents. Countries like Canada, USA, Sweden, Malaysia, etc. are considered as water rich countries and their per capita availability is above 10 cu.m/year. Counties like India, China, South Africa, etc are water poor countries and the per capita availability is less than 5 cu.m/year.

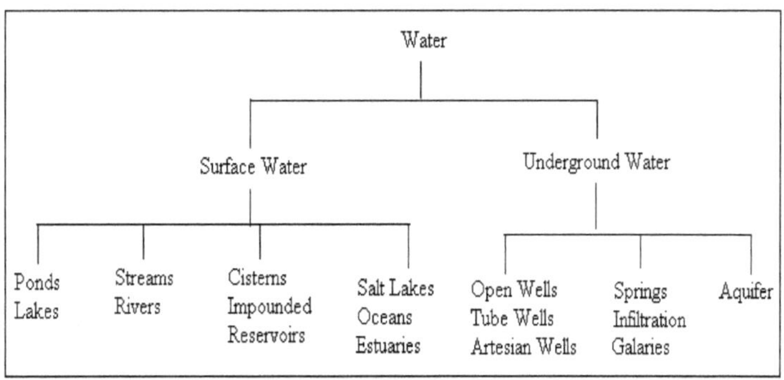

Cisterns

These are rain water storage tanks. Rain water from roofs, courtyards etc. is collected in a water tight storage tank and is used when there is an excess water demand.

Impounded Reservoirs

An impounded reservoir, also called embankment pond. It is made by building an embankment or earth fill across a narrow valley so that, a completely new surface water storage.

Water is impounded by constructing a bund, a weir or a dam.

Artesian Wells

Mostly, they are formed in the valley portion of hills. As the hydraulic gradient line passes above the mouth of the well, water comes out of the well with pressure.

Infiltration Galleries

They are horizontal conduits or pipes with perforations or open joints so that the ground water can enter them by gravity. The water is then pumped from the wet well located at the end or at the middle of the gallery. Recent times, they are extensively used as they yield good quality water throughout the year at low cost.

Aquifers

Underground water layer associated with permeable rock is called aquifer. The water quality becomes good since it is passed through the sieves of gravel, clay, crystalline rocks, etc. Aquifer literally means "a barrier". It is below 100 feet from earth's surface.

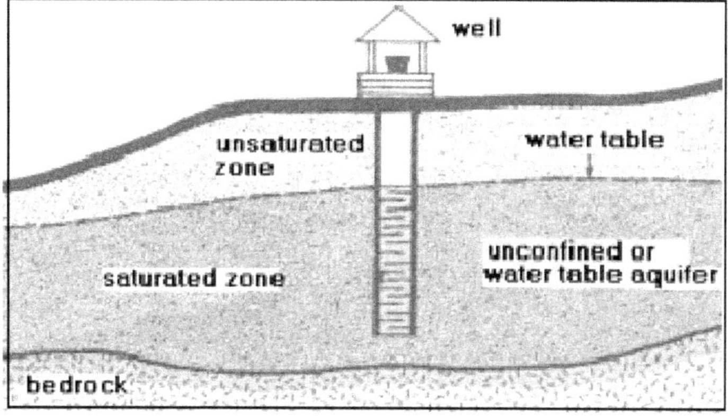

Unconfined Aquifer

Unconfined aquifers are those into which water seeps from the ground surface directly above the aquifer. It is less prone for contamination by pollutants discharge.

Confined Aquifer

Confined aquifers are those in which an impermeable dirt/rock layer exists underneath and above the aquifer.

It prevents water from seeping into the aquifer directly from the ground surface. Water enters the aquifer from far away region were impermeable layer does not exist.

Uses of Water

Water in its pure form is a best medicine as such and also a medium for many other medicines. Water is used by all biological systems and nutrient cycle passes through water. Water is an universal solvent and hence used in all type of activities both in life processes and also in industrial environment.

Water is required for drinking, cooking, bathing, washing of hands etc., flushing of urinals or water closets, washing clothes, washing floors, gardening, cleaning vehicles and maintaining domestic animals.

Use	Description
Domestic Purposes	As per the Bureau of Indian Standards, the average water demand is 135 litres per capita per day.
Commercial Purposes	The commercial activities are a function of the population of the town. The water demand for commercial purposes may be taken as 10–20% of domestic water demand or 15 to 25 litres/capita/day.
Municipal/Civic Purposes	These cleaning public lavatories, large markets, sewers etc. Further aesthetics in town by beautifying it with gardens, lawns, fountains, artificial ponds, etc.
Irrigation Purpose	Throughout the world, irrigation (water for agriculture, or growing crops) is probably the most important use of water. Almost 60% of all the world's fresh water withdrawals go towards irrigation uses.
Fire Demand	Water is the cheapest fire extinquisher
Industrial Purposes	Water demand depends on the size or capacity or process of the industry. This may be taken as 20–25% of the total demand.
Losses and Wastage	This is due to leakage of water from faulty pipe–joints and faulty valves and fittings. While estimating the water demand of a city such wastage may be taken as 15% of total demand.

Over Utilization of Surface and Ground Water

The use of water beyond sustainable limit is termed as over utilization. At present demand of water is growing by 2.4% per year.

Over utilization is due to increasing population, excessive demands due to Industrial technologies, varied and strange life activities of human beings, etc. It is estimated that by 2025 there would not be any ground water for exploitation. It is ugly to say we waste about 30% of the water to sea.

Consequences of Over Utilization of Water

- Decrease in ground water level. Sinking of overlaying land surface and structural damage
- Lowering of water table and change in the direction of underground water flow.
- Intrusion of salt water

- Earthquake and landslides
- Water pollution
- Drying up wells and loss of irrigation income.

In this direction Government of Tamil Nadu has taken active steps to preserve underground water by minimizing bore wells and also implementing mandatory requirement of rain water harvesting method in every house. In fact according to report underground water level is increased considerably after such effort.

Flood

A flood is an overflow of water, an expanse of water submerging land. In many arid regions of the world, the soil has very poor water retention characteristics, or the amount of rainfall exceeds the ground's ability to absorb water.

Sometimes rain results in a sudden flood of water filling dry streambeds known as a "flash flood".

Many rivers are a stream of flow of water over relatively flat land border plains. When heavy rainfall or melting snow causes the river's depth to increase and rapidly cover the adjacent plain.

Sl.No.	Causes	Consequences	Control
1	Monsoon rainfalls can cause disastrous flooding in some equatorial countries, such as Bangladesh, due to their extended periods of rainfall.	Soil erosion, Flood carries fertile sediment and sediment deposits to lower areas.	Reducing the scale of floods through better catchments management, controlling runoff, detention basins, dams and protection of wetlands.
2	Hurricanes, Tsunami have a number of different features, which, together, can cause devastating flooding.	Loss of materials and life. Severe flood can damage buildings, bridges, etc.	Isolation of flood threat by flood embankments, flood proofing and limiting floodplain development.
3	Closing of water streams, utilization of lakes, ponds, riverbeds etc for developmental activity, building apartments or commercial complex etc also leads to flood.	Damages the transport system, drainage, etc.	Rainwater harvesting, building dams and constructing artificial watersheds.
4	Flood occurs in general when heavy rainfall, melting ice or snow or combination of these, exceeds the carrying capacity of the receiving stream.	Prolonged flood submerges agricultural area and production decreased.	Construction of dams. Watersheds and proper drainage system. Periodical desilting of ponds, lakes and water reserviours.

National Commission of Floods estimated that land area prone to flood has doubled from 20 million hectares in 1971 to 40 million hectares in 1980. The damages cost was Rs. 21 crores in 1971 and is increased to Rs. 1130 crores in 1980.

Rought

A Drought is an extended period where water availability falls below the statistical requirements for a region.

The Sahelian drought that began in 1968 was responsible for the deaths of between 100,000 and 250,000 people, the disruption of millions of lives, and the collapse of the agricultural bases of five countries.

There are generally three types of conditions that are referred to as drought.

- **Meteorological drought** is brought about when there is a prolonged period with below average precipitation.

- **Agricultural drought** is brought about when there is insufficient moisture for average crop or range production. This condition can arise, even in times of average precipitation, due to soil conditions or agricultural techniques.

- **Hydrologic drought** is brought about when the water reserves available in sources such as aquifers, lakes, and reservoirs falls below the statistical average. This condition can arise, even in times of average (or above average) precipitation, when increased usage of water diminishes the reserves.

- **Socioecomic drought** is characterized by poor water quality and quantity leading to health problems. It leads to economic loss and lie loss.

When the word "drought" is used commonly, the most often intended definition is meteorological drought. However, when urban planners use the word, it is more frequently in the sense of hydrologic drought.

Sl.No.	Causes	Consequences	Control
1	Ecological balance is lost. Can not maintain different seasons of the year.	Land become degraded leading desertification	Cultivation of draught resistant plants.
2	Changes in agricultural pattern and industrialisation.	Lack food leading to famine	Water Recycling
3	Deforestation and urbanisation.	Economic Loss due to lack of productivity	Water Conservation
4	Environmental pollution namely by vehicular emissions.	It will not support life forms and biodiversity is lost in the area.	Sustainable Use of Water, aforestation

Israel uses modern agricultural methods to overcome the drought. Similarly, such practices are suggested in India too. Social forestry is an alternate method proved to cope up with drought. Kolar and Belagum in Karnataka uses social forestry method to improve draught affected areas. Government of India is taking active steps in linking rivers to create a sustainable development in draught affected areas, especially in Deccan Peninsula. Israel uses seawater desalination method for its water requirements.

Conflict Over Water

In fact, there is a long and highly informative history of conflicts and tensions over water resources, the use of water systems as weapons during war, and the targeting of water systems during conflicts caused by other factors. The current categories, or types of conflict, now include:

- **Control of Water Resources** (state and non-state actors): where water supplies or access to water is at the root of tensions.
- **Military Tool** (state actors): where water resources, or water systems themselves, are used by a nation or state as a weapon during a military action.
- **Political Tool** (state and non-state actors): where water resources, or water systems themselves, are used by a nation, state, or non-state actor for a political goal.
- **Terrorism** (non-state actors): where water resources, or water systems, are either targets or tools of violence or coercion by non-state actors.
- **Military Target** (state actors): where water resource systems are targets of military actions by nations or states.
- **Development Disputes** (state and non-state actors): where water resources or water systems are a major source of contention and dispute in the context of economic and social development.

It is evolving as international and regional politics and as new factors become increasingly, or decreasingly, important in the affairs of humanity. In all this, however, one factor remains constant: the importance of water to life means that providing for water needs and demands will never be free of politics.

Water and International Conflict

The scarcity of water for human and ecosystem use leads to intense political pressures, often referred to as "water stress." As a consequence, competition for water resources has contributed to tensions around the globe between competing users.

Examples of internal water conflicts range from interstate violence and death along the Cauvery River in India. The trend is extended to California farmers blowing up a pipeline meant for Los Angeles, to much of the violent history in the Americas between indigenous peoples and European settlers.

Development, Crisis and Conflict Resolution

- Unilateral decisions such as Egypt's plans for a high dam on the Nile or Indian diversions of the Ganges to protect the port of Calcutta etc.
- Poor relation between water sharing States.
- Military conflict.

In fact, the last (and only) war fought specifically over water took place 4,500 years ago, between the city-states of Lagash and Umma along the Tigris River.

Over the last 50 years, there have been only 37 acute disputes (those involving violence). During the same period, 157 treaties were negotiated and signed.

For example, the Mekong Committee has functioned since 1957, exchanging data throughout the Vietnam War. Secret "picnic table" talks have been held between Israel and Jordan ever since the unsuccessful Johnston negotiations of 1953-55.

And the Indus River Commission not only survived two wars between India and Pakistan and look after the conflict between the riparian States.

Case Study

Kaveri (Krishna River) Dispute

The Cauvery River in India is a classic case of upstream-downstream country within one country, albeit between two states. The upstream Karnataka and downstream Tamil Nadu fight for water. Karnataka built Krishnaraja Sager dam across Cauvery. Once in a while releases water to Tamil Nadu. Tough Cavery Water Tribunal Award is binding on both State. They also fought in the Supreme Court the order is not followed properly.

The Satluj-Yamuna Link (SYL) Dispute

It is the dispute between States of Punjab and Haryana regarding water sharing of rivers Ravi and Beas. The Eradi Tribunal was constructed by Supreme Court of India (1985). Though the sincere efforts are made by the committee yet to resolve the difference between states.

However according to Indian Constitution, River is a National Property and States have limited rights and thus needs wider look by the states concerned in the dispute.

Sardar Sarovar Project

The World Bank's withdrawal from the Sardar Sarovar Project in India in 1993 was a result of the demands of local people threatened with the loss of their livelihoods and homes in the submergence area. This dam in Gujarat on the Narmada has displaced thousands of tribal folk, whose lives and livelihoods were linked to the river, the forests and their agricultural lands. While they and the fishermen at the estuary, have lost their homeland, rich farmers downstream will get water for agriculture.

It is a question of social and economic equity as well as the enormous environmental losses, including loss of the biological diversity of the inundated forests in the Narmada valley.

The dispute continued though no solution achieved. The project is under way inspite of many criticisms. Indefinite hunger strike has been called off by the request of honorable President of India Dr. A.B.J. Abdul Kalam.

Dispute Resolution-Indian Scenario

Sarkaria Commission in its report has recommended amendments in legal framework on Interstate River Water Disputes.

1. Union Govt. to constitute a Tribunal with one year from date of receipt of application of any disputant state.
2. Award of Tribunal to become effective within five year from date of constitution of a tribunal.
3. Tribunal's award should have the same force and sanction. It is as an order or decree of Supreme Court to make a tribunal's award really binding.

Treaty

This is an agreement between water sharing parties by amicable settlement without intervention of court.

The Indus Water Treaty (1960)

It is the treaty between India and Pakistan regarding water sharing with respect to river Indus. Accordingly, the tributaries Indus, Jhelum and Chenab were allotted to Pakistan, Satlug, Ravi and Beas were allotted to India.

Dams Benefits and Problems

Since independence, it was the dream of Pandit Jawaharlal Nehru, then Prime Minister, India to build dams for self sufficiency in food and power, and hence about 700 dams primarily to store water during monsoon time.

In this about 90% water is used for irrigation purpose. By 2005, India hopes to provide water for about 113 million hectares. Dams have its own benefits and drawbacks.

Sl.No	Benefit	Problems
1	providing inexpensive electricity	Submerging land surface.
2	flood control	Displacement of people in upstream and nearby downstream area
3	recreation	Water logging and spread of diseases
4	irrigation	Increased water pressure on the earth leading to earthquake
5	industry growth	Conflict over water sharing
6	navigation	Failure of dam leading to life and property loss
7	Water Security, reservoir when monsoon fails	Capital intensive and benefit arising out of it may not be adequate to expenditure.

Displacement of People

Estimates suggest that worldwide, dams have displaced some 40 million to 80 million people, yet mitigation, compensation and resettlement attempts are often inadequate. The Three Gorges Dam on the Yangtze River, due for completion in 2009, will displace 1.3 ~ 1.8 million people, 320 villages and 140 towns along with very many ancient and historic sites.

Problems of Rehabilitation

- Rehabilitation is always considered to be an obstacle rather than enforcing the right of the country men.
- No guidelines and even if one which not implemented.
- Ecological niche of the tribal peoples are steeled without adequate consideration or compensation.
- Illiteracy and inability of voice is taken advantage.
- False promises and cheating are common among project implementers and contractors.
- Tribal people cannot understand language of court both in verbal and script.
- Voluntary organization and environmental activists have their own limitations.
- Resettlement is a general slogan to cool tribal people and in legal language.

Rehabilitation Policy

- The displaced people should get the share of fruits from development.
- Should create environment equivalent to the lost one.
- Poverty eradication should be coupled with rehabilitation.
- Popper should be given reasonable employment.
- Tribals should be well informed their rights and be put in their line of life.
- The right on their land and forest is maintained even after displacement.

- They should be administered by the team of their own people.
- Adequate compensation and facilities for shifting and also the requirement for their initial period life (about 6 months).

Decision to Build a Dam

Before any decision is taken to build a dam for whatever purposes a systematic analysis is required without eternal or political pressure.

i. **Agriculture and Irrigation:** About 20% of the world's agricultural land is irrigated, but this land produces 40% of the world's food. Thus water and its sustainable use is important to consider while building a dam.

ii. **Energy:** It is the yet another use of water in dams. Hydrothermal energy is green energy. Thus total energy production, its economic factor, etc, to be considered while constructing a dam.

iii. **Water Supply:** Dam water is also used for municipal supply for drinking and other domestic purposes. The water required and it benefit should be taken into consideration.

iv. **Flood Management:** In flood prone areas, dams if no other alternative give a solution to flood management. The water trapped in dams can be used for economic purposes.

Mineral Resources

A mineral is a naturally occurring substance of definite chemical composition and identifiable physical properties. An ore is a mineral or combination of minerals from which a useful substance, such as a metal, can be extracted. Minerals are non-renewable resource. It has localized occurrence and present in limited quantity. The quality also varies from place to place. The presence of coal and petroleum is bigger compared to metallic deposits.

Use and Exploitation of Mineral Resources

The demand for minerals is increasing with increasing technological improvements and life styles. Further population growth supplements the demand for minerals. The per capita consumption of mineral is rising.

- All technological process involves metals and non-metal, which are the main products of minerals. All industrial machines, process, equipments, etc includes minerals.
- Many minerals are having medical properties and used as such or as compounds in chemotherapy treatments.
- Minerals are also used as energy resources as in the case of nuclear energy production.

- It is used in construction of houses, transport equipments, etc.

- It is used as ornaments and aesthetic purposes.

- Many minerals like coal, lignite, etc are used as fuels.

- Economic wealth of a country is judged by its abundance in mineral resources.

- Communication system depends on minerals.

Indian Mineral Resource Scenario

India produces 84 minerals, including 4 fuel minerals. She is self sufficient in 30 minerals, which includes, coal, bauxite, iron and manganese. India is fifth largest producer of coal and bauxite. There is appreciable quantity of copper, gold, iron, diamond, lead and zinc in Indian sub-continent. India also blessed with seabed mineral resources.

Government of India has recognized the importance of exploration of oil and mineral resources. The few mineral exploration companies are

(i) Mineral Exploration Corporation Limited (MECL).

(ii) National Aluminum Company Limited (NALCO).

(iii) Malanjkhand Copper Project.

(iv) Femnor Mineral India.

(v) Bharat Aluminum Company Limited.

(vi) Bharat Gold Mine Limited.

(vii) Hindustan Zinc Limited.

(viii) Hindustan Copper Limited.

(ix) Sikkim Mining Corporation.

India has bilateral cooperation with France, China, Australia, Russia and Vietnam to assist the exploration and technological innovation in mining, operation and conservation of minerals. Recent years foreign companies also encouraged to carry out mining operations in India.

Environmental Effects of Extraction and Use of Mineral Resources

- Mineral extraction requires enormous amount of energy and water, causes ecological disturbance.

- Mineral extraction pollutes the nearby water resources and also air pollution due to the emission of green house gases.

- The extraction process generates side products and wastages, which causes serious concerns in term of disposal and safety.

- It leads to industrialization and urbanization which is not welcomed in terms of ecology and environmentalogy.
- Huge industrial process causes deforestation, rehabilitation, loss of fertile land etc, due to construction of transport paths and bridges.
- Increasing wildlife disturbance and mortality, due to noise, accidents, decrease of vegetation etc.
- Large scale use toxic chemicals and uncontrolled spillage causes the spread of diseases.
- Industrial accidents is a common concern.

Case Studies

Mining in Udaipur

In Udaipur illegal mining and quarrying of phosphate rocks, dolomite and building stones. The uncontrolled use explosives causes air and water pollution. It adversely affects irrigation and also migration of wildlife.

Quarrying Rare Earths in Kanyakumari District

Indian Rare Earths Corporation is quarrying sands near the sea shore of Manali, Kanaykumari District, which rich in Thorium and Uranium for the purpose of foreign currency. It is leading to the loss of many coconut plantation and sea shore beauty. Further many people lost their habitat.

Food Resources

Today our food comes almost entirely from agriculture, animal husbandry and fishing. Although India is self-sufficient in food production, it is only because of environment with excessive use of fertilizers and pesticides.

Among 30,000 plant species, only 15 plants and 8 animal species supply about 90% of food.

World Food Problems

In many developing countries where populations are expanding rapidly, the production of food is unable to keep pace with the growing demand.

Food production in 64 of the 105 developing countries is lagging behind their population growth levels. These countries are unable to produce more food, or do not have the financial means to import it. However, the global food production is increased substantially, the hunger is yet to be eliminated. This is because poor has not got enough land and money. Research reveals that many countries including India food scarcity is not due to lack of food rather due to uneven distribution among different class of people.

Environmental degradation, soil erosion, water lagging, desertification. Pollution, etc causes serious problem in food sufficiency in coming years. Agricultural revolution (Green Revolution) had made progress in mechanisation of agriculture and thus productivity, but not in phase with population increase.

Respectfully October 16 is marked for World Food Day. Still statistically every one out of four is malnourished or undernourished.

India and Food Security

India is one of the countries that have been able to produce enough food by cultivating a large proportion of its arable land through irrigation. The Green Revolution of the 60's reduced starvation in the country.

However many of the technologies we have used to achieve this is now being questioned. In India 40% (50% agricultural workers) suffer from malnutrition mostly children between age 1-5 and women (mostly pregnant women). Change of food habit, hunger death is also reported from India due to drought, famine, etc.

The developed countries also have poverty malnutrition and hunger. According to one estimate in USA 20 million people (12 million children and 8 million adults) are suffering from chronic hunger and malnutrition. The main reason is cultural, social and managerial problems not because of lake of food availability.

Changes Caused by Modern Agriculture

Modern agriculture is said to be scientific one. It proposed to use fertilizers, pesticides, herbicides, etc. It is aimed at better productivity. Many better yielding genetically modified crops are under cultivation.

Technology driven agriculture is modern one. Thus tractors, harvesting machines are under use in recent days. Newer methods like drip irrigation, tube well water supply etc are adopted. Crop rotation is also practiced. Indeed modern agriculture made formers become technocrat. On the other hand many formers become workless under landlords. More and more become industrial worker. The present trend is younger generation is not behind agriculture.

Loss of genetic diversity in crop plants is another issue that is leading to a fall in agricultural products. Rice, wheat and corn are the staple foods of two thirds of the world's people.

As wild relatives of crop plants in the world's grasslands, wetlands and other natural habitats are being lost, the ability to enhance traits that are resistant to diseases, salinity, etc. is lost. Genetic engineering is an untried and risky alternative to traditional cross breeding.

High yielding verities are monocultured and hence soil losses its fertility irreversibly over a period.

Effects of Modern Agriculture

Since 1950 the increase in global food production was the result of increased yield per unit area and the achievement is called as green revolution. The process involves the following steps.

(i) Selective breed or genetically engineered high yield and resistant verities are selected for culture.

(ii) Excessive use of fertilizers, pesticides and water. Salinization and water logging has affected a large amount of agricultural land worldwide.

(iii) Decreasing cropping time with 3 to 4 crops per year.

The first green revolution took place between 1950 and 1970. The second green revolution started since 1967. The net effects are as follows.

- The yield increased from 2 to 5 times in a given area of land.
- Increased use of fossil fuels as the number of tractors used in agriculture area increased.
- Commercial fertilizers have raised 10 fold.
- The pesticide use has increased 10 times.
- Intensive agriculture, combination with fossil fuel and cultivation of leguminous crops led to huge increase of nitrogen deposits in terrestrial and aquatic ecosystem.
- Overexploitation of land leads to soil erosion, lose of fertility, salinity, water logging etc.
- Water and land get polluted with pesticides, nitrates and hence not usable.

Pesticide Problems

As with fertilizers, application rates for chemical pesticides are highest in industrialized countries, reflecting the low cost of farm inputs in relation to farm incomes. Application rates in Vietnam, for example, are around 1.0 kg/ha, compared to 11.8 kg/ha in Korea and 19.4 kg/ha in Japan.

Endosulfan is an endocrine disrupter and genotoxic, attacking the central nervous system, kidneys, skin and reproductive system. For over 20 years villagers in Kasargod district, with some 4,500 hectares of cashew plantations, have been affected by an unusually large number of cancer deaths, neurological disorders and physical and mental impairments. It is attributed due to the use of Endosulfan.

Pesticide production Japan and India are important producers of technical grade pesticides. Korea and Taiwan also produce technical grades.

Moreover such high doses of pesticides generate resistance among the pests over time and cause more damage to the beneficial organisms.

Use of predators will reduce the excessive use of pesticides and related problems.

Biological Amplification

The process of accumulation of pesticides in cells of animal is known as biological amplification. When an animal takes up a pesticide, it is accumulated in the fat tissues. If a carnivore eats the above it is further concentrated in the latter. The consequence is even if the lower animal is not affected, the higher animal might die due to pesticide poison.

A small amount of DDT is sprayed in aquatic ecosystem, it accumulates up to 250 times in the small aquatic organisms, like shrimp, clams, etc. Larger aquatic organisms eat the smaller organisms and the DDT concentration increases as high as 2000 times. When birds consume fish it is further biologically amplified and may cause death. It is reported that DDT interferes with the production of eggshells and makes it fragile

Case Study

DDT (Dichloro Diphenyl Trichloroethane) used in the United States before 1966 may have caused an epidemic that has only now been detected! According to a fascinating study published in the July 14, 2001 issue of The Lancet, scientists who studied stored cord blood samples from mothers who had delivered at that time found elevated levels of DDT breakdown products among the group who had premature deliveries or low birth weight infants. This would make DDT responsible for a host of medical problems and the deaths of many children. DDT use in the United States was stopped in 1972 because it caused reproductive damage to birds. DDT is still widely used in 25 countries, including India for insect control.

Pepsi Case

Food Centre for Science and Environment (CSE) India has found that Coke and Pepsi are selling soft drinks with pesticide content 30-40 times higher than EU guidelines. The centre said the soft drinks in India had high pesticide residues because the soft drink and bottled water industry uses an enormous amount of ground water as the basic raw material. report said concentration of lindane, a carcinogen that also damages the nervous system.

Kerela State Pollution Control Board finding high levels of cadmium in the waste from Coke's Pallakad plant and classifying it as Hazardous waste.

Arguments

DDT is almost completely harmless in humans, even in doses many times larger than what's found in Indian water. Numerous studies have been performed showing with high confidence that it is not even slightly correlated with cancer.

Indeed, the original reason DDT became popular in the 1940's was because it was found to be dramatically safer than any other pesticide. During World War II was commonly applied directly to humans as a delousing method, and even as a mosquito repellant.

In small doses, they may be almost completely harmless, though. It's kind of like cyanide - we need it in order to have any vitamin B12, and it's contained in a wide variety of good food - cherries, almonds, nutmeg, lima beans, and of course Cassava. But in huge quantities it overwhelms your body's cyanide-processing enzyme (rhodanase), and that's when it can kill you.

Accordingly, it appears DDT is legally considered to be a vitamin supplement and essential as like cyanide for Vitamin B12. So allowed continue sell in the Indian market in such limit for Indians.

Fertilizer Use and Problems

Only 30-35% fertilizer is utilized by the crop, the remaining proportion reaching the underground water resources. It leads to micronutrient imbalance in the soil leading to loss of fertility of land.

Often leads to contamination of ground water by fertilizer. Salinity and TDS (Total Dissolved Solid) increased leading to health problem to living beings.

Hardness and alkalinity of water also will be high, high enough to classify as wastewater. Micronutrient imbalance, in which soil looses its micronutrients and it becomes infertile. Nitrate pollution due to excess and uncontrolled use leading to health hazards like "Blue Baby Syndrome or Methaemoglobinemia", a type of anemia prevalent in England, France, Germany, etc. Eutrophication occurs due to the presence of large amount of nitrogen and phosphorus in the water. Health hazards and spread of diseases is also anticipated.

Eutrophication

Eutrophication

Eutrophication is a process in which excess nutrients stimulate excessive plant growth (algae, periphyton attached algae, and nuisance plants weeds). This enhanced plant growth, often called an algal bloom. It reduces dissolved oxygen in the water when dead plant material decomposes and can cause other organisms to die. It also crate bad odour to water body.

Nutrients can come from many sources, such as fertilizers applied to agricultural fields, golf courses, and suburban lawns; deposition of nitrogen from the atmosphere; erosion of soil containing nutrients; and sewage treatment plant discharges. Water with a low concentration of dissolved oxygen is called hypoxic.

Alternatives to Fertilizers

- Organic farming is gaining importance and appears to be one alternative.
- Avoid indiscriminate use of fertilizers. Use the calculated amount.
- Encourage traditional methods with supplement of modern techniques.
- Vermiculture is also an alternative method for avoiding fertilizer problems.

Organic Farming

As organic farming refrains from using chemical inputs, the health hazards and other side effects posed by those inputs are prevented. In fact, the first 'scientific' approach to organic farming can be quoted back to the Vedas of the 'Later Vedic Period', the essence of which is to live in partnership with, rather than exploit, nature. More recently, Mahatma Gandhi pioneered organic agriculture through his constructive programmers in several locations in India.

It is estimated that only around 30% of the total cultivable area in India, where irrigation facilities are available, is covered under green revolution technology. The remaining 70% (approximately) of arable land, where rainfed or dry farming is practiced, do not, generally, use agro-chemicals.

Water Logging

Water logging is when the soil surface area becomes saturated and remains so for year(s). Soil pores (spaces) are full of water. Excess water cannot drain away. In India nearly 46 million hectares of water from various water bodies like river, streams, lakes etc, irrigating nearly 170 million hectares of land.

Sl.No.	Causes	Consequences	Control
1	Periods of heavy rain.	Deceased oxygen in the soil.	Management of drainage lines for efficient water flow.
2	Poor irrigation management.	Vegetation can turn yellow, growth is stunted and thin	Management of surface water-flow to avoid surface erosion.
3	Poor drainage	Trees and plants can die.	Apply only as much as the plant will use.
4	Over-watering with irrigation causes water logging.	Bare patches of soil appear on crops appear and poor quality of products	Increase deep rooting vegetation for greater utilisation of water from the soil.
5		Plants tolerant to saturated conditions will take over.	Cultivate crops and plants that can survive in excess water.

Salinity

The presence of excessive salts in soil is salinity. The salts of sodium, magnesium, calcium, etc are accumulated. The pH of the soil is also alkaline. It is estimated that 10 million hectares are now being lost every year as a result of salinity.

Sl.No.	Causes	Consequences	Control
1	Use of water with large amount of salts.	Poor growth of crop and decreased yield with poor quality.	Changes in cultivation to salt tolerant and utilizing crops.
2	Excessive use of fertilizers.	Poor disease and pests resistance.	Use calculated amount of fertilizers.
3	Excessive use of water and water logging	The land soon becomes wasteland.	Cleans up the salt content by scientific methods like treating with EDTA
4	Discharge of effluents with high slat concentration		Use better standard water for irrigation.
5			Use better drainage system and avoid water logging

Energy Resources

Physicists define energy as the capacity to do work. Energy is found on our planet in a variety of forms, some of which are immediately useful to do work, while others require a process of transformation.

The sun is the primary energy source in our lives. We use it directly for its warmth and through various natural processes that provide us with food, water, fuel and shelter.

Chemical energy, contained in chemical compounds is released when they are broken down by animals in the presence of oxygen. Electrical energy produced in several ways, powers transport, artificial lighting, agriculture and industry. Nuclear energy is held in the nucleus of an atom and is now harnessed to develop electrical energy. We use energy for household use, agriculture and production of industrial goods and for running transport.

Growing Energy Needs

For almost 200 years, coal was the primary energy source fuelling the industrial revolution in the 19th century. At the close of the 20th century, oil accounted for 39% of the world's commercial energy consumption, followed by coal (24%) and natural gas (24%), while nuclear (7%) and hydro-renewables (6%) accounted for the rest.

At present almost 2 billion people worldwide have no access to electricity at all. While more people will require electrical energy, those who do have access to it continue to increase their individual requirements.

Energy has always been closely linked to man's economic growth and development. Present strategies for development that have focused on rapid economic growth have used energy utilization as an index of economic development.

Between 1900 to 2000 world energy consumption has increased by a factor of fourteen times but the population increased by only three times.

Year	Population(Billions)	Fuel Consumption (Billion tons of Coal equivalent)
1900	1.6	1
1950	2.9	3
2000	6	14

In 1998, the World Resources Institute found that the average American uses 24 times the energy used by an Indian.

Between 1950 and 1990, the world's energy needs increased four fold. The world's demand for electricity has doubled over the last 22 years! Electricity is at present the fastest growing form of end-use energy worldwide.

Indian Energy Production Profile

As per the 1995 calculation, in India coal accounted 63.3 %, petroleum accounted 18.6 %, hydroelectricity 8.9 %, natural gas 8.2 % and nuclear power 1 %.

Forecast reveals India will be energy deficient country by 30 % and dependent on energy import. India's current energy production is about 9 Btu and expected to be 16.4 Btu in 2010. In India, biomass (mainly wood and dung) accounts for almost 40% of primary energy supply. While coal continues to remain the dominant fuel for electricity generation, nuclear power has been increasingly used since the 1970s and 1980s and the use of natural gas has increased rapidly in the 80s and 90s.

India depends mostly on fossil fuel and polluting natural environment. India is 7th largest consumer of coal and largely used in thermal power plants. Gasoline and diesel are the primary fuels used in vehicles.

Poor standard of fuel and inefficient combustion technologies further contributes to environmental pollution.

Types of Energy

There are three main types of energy. They are:

- Non-renewable energy
- Renewable energy
- Nuclear energy

However, this classification is inaccurate because several of the renewable sources, if not used 'sustainably', can be depleted more quickly than they can be renewed.

Non Renewable Energy

To produce electricity from non-renewable resources the material must be ignited. The fuel is placed in a well contained area and set on fire. The heat generated turns water to steam, which moves through pipes, to turn the blades of a turbine. This converts magnetism into electricity, which we use in various appliances.

Non-Renewable Energy Sources

These consist of the mineral based hydrocarbon fuels coal, oil and natural gas, that were formed from ancient prehistoric forests. These are called 'fossil fuels' because they are formed after plant life is fossilized. At the present rate of extraction there is enough coal for a long time to come. Oil and gas resources however are likely to be used up within the next 50 years.

Renewable Energy

Renewable energy systems use resources that are constantly replaced and are usually less polluting. Examples include hydropower, solar, wind, and geothermal (energy from the heat inside the earth). We also get renewable energy from burning trees and even garbage as fuel and processing other plants into biofuels.

One day, all our homes may get their energy from the sun or the wind. Your car's gas tank will use biofuel. Your garbage might contribute to your city's energy supply.

Renewable energy technologies will improve the efficiency and cost of energy systems. We may reach the point when we may no longer rely mostly on fossil fuel energy.

Use of Alternate Energy Resources

Hydroelectric Energy	Biogas Energy
Solar Energy	Wind Energy
Photovoltaic Energy	Tidal and Wave Energy
Mirror Energy	Geothermal Energy
Biomass Energy	Nuclear Energy

Hydroelectric Energy

The principle is water flowing down by a natural gradient to turns turbine to generate electricity known as 'hydroelectric energy/power'. Between 1950 and 1970, Hydropower generation is increased worldwide.

In 1882, the first Hydroelectric power dam was built in Appleton, Wisconsin. In India the first hydroelectric power was built in the late 1800s and early 1900s by the Tatas, in the Western Ghats of Maharashtra. Jamshedjee Tata, a great visionary who developed industry in India.

Hydroelectricity is globally important as it provides 19% of the world's electricity supply. However, it becomes even more important when examined at a national scale, as hydropower provides over 90% of national supply in 24 countries, and over 50% of national supply in 63 countries.

Advantages

The main advantage of hydropower is that it is renewable and has low running costs.

Water is required for many other purposes besides power generation. These include domestic requirements, growing agricultural crops and for industry.

It is possible to produce hydroelectric power without a large dam and reservoir. Run-of-the-river schemes (generates energy as the water flows) can provide a lot of the benefit at far lower cost.

Drawbacks

- To produce hydroelectric power, large areas of forest and agricultural lands are submerged. These lands traditionally provided a livelihood for local tribal people and farmers.
- The use of rivers for navigation and fisheries becomes difficult once the water is dammed for generation of electricity.

- Resettlement of displaced persons is a problem for which there is no ready solution.

- In certain regions large dams can induce seismic activity, which will result in earthquakes.

Solar Energy

In one hour, the sun pours as much energy onto the earth as we use in a whole year. If it were possible to harness this colossal quantum of energy, humanity would need no other source of energy. The energy received near earth space is aout 1.4 kJ/sec/m^2 (solar constant). Solar cells (photovoltaic cells) are used to harvest solar energy.

Solar Heating for Homes

A passive solar home or building is designed to collect the sun's heat through large, south-facing glass windows. In energy efficient architecture the sun, water and wind are used to heat a building when the weather is cold and to cool it in summer.

Solar Water Heating

Most solar water-heating systems have two main parts: the solar collector and the storage tank. The solar energy collector heats the water, which then flows to a well insulated storage tank. A common type of collector is the flat-plate collector, a rectangular box with a transparent cover that faces the sun, usually mounted on the roof. Small tubes run through the box, carrying the water or other fluid, such as antifreeze, to be heated. The tubes are mounted on a metal absorber plate, which is painted black to absorb the sun's heat. The back and sides of the box are insulated to hold in the heat. Heat builds up in the collector, and as the fluid passes through the tubes, it too heats up.

Solar Cookers

The heat produced by the sun can be directly used for cooking using solar cookers. A solar cooker is a metal box, which is black on the inside to absorb and retain heat. The lid has a reflective surface to reflect the heat from the sun into the box. The box contains black vessels in which the food to be cooked is placed.

Other Solar-Powered Devices

Solar desalination systems (for converting saline or brackish water into pure distilled water) have been developed. Solar cars, solar traffic signals, solar street lights, solar satellites, etc are present generation applications. In future, they should become important alternatives for man's future economic growth.

Advantages of Solar Energy

- No or limited environmental pollution.
- Higher safety than many other energy harvesting techniques.
- Cheap and readily available for billions of years.
- No emission of side peoducts.

Disadvantages of Solar Energy

- There should be sun shine atleast few hours in most of the days.
- Low efficiency of trapping, better technology required.
- Storage and back up devices (batteries, capacitors) required which are costly and not environmentally friendly.
- Converters are required for DC to AC conversion.

Biomass Energy

Biomass energy is a form of stored solar energy. Although wood is the largest source of biomass energy, we also use agricultural waste, sugarcane wastes, and other farm byproducts to make energy. There are three ways to use biomass. It can be burned to produce heat and electricity, changed to a gas-like fuel such as methane, or changed to a liquid fuel.

Note that like any fuel, biomass creates some pollutants, including carbon dioxide, when burned or converted into energy. In terms of air pollutants, biomass generates less relative to fossil fuels.

Biogas

Biogas is produced from plant material and animal waste, garbage, waste from households and some types of industrial wastes, such as fish processing, dairies, and sewage treatment plants. It is a mixture of gases, which includes methane, carbon dioxide, hydrogen sulphide and water vapour.

Denmark produces a large quantity of biogas from waste and produces 15,000 megawatts of electricity from 15 farmers' cooperatives.

London has a plant, which makes 30 megawatts of electricity a year from 420,000 tons of municipal waste which gives power to 50,000 families. In Germany, 25% of landfills for garbage produce power from biogas. For example Japan uses 85% and France about 50% of its waste for biogas production.

In India in the rural sector, the biogas plants use cow dung, which is converted into a gas, which is used as a fuel. It is also used for running dual fuel engines.

The fibrous waste of the sugar industry is the world's largest potential source of biomass energy. Ethanol produced from sugarcane molasses is a good automobile fuel and is now used in a third of the vehicles in Brazil.

The National Project on Biogas Development (NPBD), and Community/Institutional Biogas Plant Program promote various biogas projects.

Wind Energy/Power

Most of the early work on generating electricity from wind was carried out in Denmark, at the end of the last century. Today, Denmark and California have large wind turbine cooperatives which, sell electricity to the government grid. In Tamil Nadu, there are large wind farms producing 850 megawatts of electricity. At present, India is the third largest wind energy producer in the world.

Sl. No	Advantages	Disadvantages
1	Cheap and easy to harvest	Steady wind flow is required
2	No air pollution	Large area of land surface is used
3	High effeciency	Energy back up system is required.
4	Easy construction	Damage caused during irregular and strong winds

Tidal and Wave Energy/Power

The energy of waves in the sea that crash on the land of all the continents is estimated at 2 to 3 million megawatts of energy. From the 1970s several countries have been experimenting with technology to harness the kinetic energy of the ocean to generate electricity. Tidal power is trapped by placing a barrage across an estuary and forcing the tidal flow to pass through turbines.

Tidal power stations bring about major ecological changes in the sensitive ecosystem of coastal regions and can destroy the habitats and nesting places of water birds and interfere with fisheries. A tidal power station at the mouth of a river blocks the flow of polluted water into the sea, thereby creating health and pollution hazards in the estuary.

Another developing concept harnesses energy due to the differences in temperature between the warm upper layers of the ocean and the cold deep sea water. These plants are known as Ocean Thermal Energy Conversion (OTEC).

Sl. No	Advantages	Disadvantages
1	Readily available thought the years.	High cost of construction
2	Pollution free	High cost of maintenance.
3	Does not depend on precipitation pattern	Damaged by costal calamities.

Geothermal Energy

It is the energy stored within the earth ("geo" for earth and "thermal" for heat). Geothermal energy starts with hot, molten rock (called magma) deep inside the earth. This is called direct use of geothermal energy, and it provides a steady stream of hot water that is pumped to the earth's surface.

In the 20th century geothermal energy has been harnessed on a large scale for space heating, industrial use and electricity production, especially in USA, Iceland, Japan and New Zealand.

Geothermal energy is nearly as cheap as hydropower and will thus be increasingly utilised in future. However, water from geothermal reservoirs often contains minerals that are corrosive and polluting.

Sl.No	Advantages	Disadvantages
1	It is efficient process.	There will be emission of Carbon dioxide
2	Low emission and environmental pollution.	Depleted is used unsustainably.
3	Low use of land surface.	Hot salt are released along with water causes environmental problems.

Nuclear Energy/Power

In 1938 two German scientists Otto Hahn and Fritz Strassman demonstrated nuclear fission of uranium by bombarding with neutrons. As the nucleus split, some mass was converted to energy. Energy released from 1kg of Uranium 235 is equivalent to that produced by burning 3,000 tons of coal.

Dr. Homi Bhabha was the father of Nuclear Power development in India. The Bhabha Atomic Research Center in Mumbai studies and develops modern nuclear technology. India has 10 nuclear reactors at 5 nuclear power stations that produce 2% of India's electricity. These are located in Maharashtra (Tarapur), Rajasthan, Tamil Nadu, Uttar Pradesh and Gujrat.

India has uranium from mines in Bihar. There are deposits of thorium in Kerala and Tamil Nadu. However India at present depends on other countries for nuclear fuel requirements.

Principle of Nuclear Energy Harvesting

Nuclear fission and nuclear fusion are the two methods by which nuclear energy is harvested.

$$U_{92}^{235} + n_0^1 \longrightarrow Ba_{56}^{139} + Kr_{36}^{94} + n_0^1 + Energy$$

The three neutrons generated further initiates nuclear fission thereby chain reaction occurs with the liberation of large amount of energy.

During nuclear fusion two nuclei fuse together gives a heavier nucleus with the liberation of energy. It is believed that sun's energy is due to nuclear fusion reaction.

$$4H_1^1 \longrightarrow He_2^4 + 2e_1^0 + Energy$$

Sl. No	Advantages	Disadvantages
1	Huge amount of energy can be generated.	Radiation hazard, waste disposal problems
2	Cheap energy source	Nuclear reactor accidents
3	Relatively lower maintenance cost.	Misuse to make atom bomb.

Mirror energy

During the 1980s, a major solar thermal electrical generation unit was built in California, containing 700 parabolic mirrors, each with 24 reflectors, 1.5 meters in diameter, which focused the sun's energy to produce steam to generate electricity.

Indian Nuclear Energy Concepts

India needs to rapidly move into a policy to reduce energy needs and use cleaner energy production technologies. A shift to alternate energy use and renewable energy sources that are used judiciously and equitably would bring about environmentally friendly and sustainable lifestyles.

India must reduce its dependency on imported oil. At present we are under-utilizing our natural gas resources. We could develop thousands of mini dams to generate electricity. India wastes great amounts of electricity during transmission. Fuel wood plantations need to be enhanced and management through Joint Forestry Management (JFM) has a great promise for the future.

Non-Renewable Energy Resources

Coal: It is a rock under the earth. It is believed that coal is formed from plants went inside the earth millions of years ago. They subjected to high temperature and pressure in anaerobic condition leading to the formation of coal. Coal is mostly carbon. It is graded based on carbon content. India has about 5% of world coal content. It is poor quality.

Petroleum: It is a liquid fossil fuel. It is believed that petroleum is formed by submerging large vegetation and animals in millions of years ago. Petroleum contain hydrocarbons. Petroleum is refined in refineries by fractional distillation.

Advantages of Non-Renewable Energy Sources

(i)   Low cost and high net energy content.
(ii) Well developed combustion technology.
(iii) Efficient distribution system.
(iv) Its used is well agreed by international community.

Disadvantages of Non-Renewable Energy Sources

(a) High environmental pollution.
(b) Last for another few decades.
(c) Mining is required to explore it and accidents are common in such process.

Land Resources

Landforms such as hills, valleys, plains, river basins and wetlands include different resource generating areas that the people living in them depend on. Many traditional farming societies had ways of preserving areas from which they used resources. The percentage of arable land in the world is about 21 %.

Land as a Resource

"Land is a delineable area of the earth's terrestrial surface. Further it includes those of the near-surface climate, the soil and terrain forms, the surface hydrology (including shallow lakes, rivers, marshes, and swamps), the near-surface sedimentary layers and associated groundwater reserve, the plant and animal populations. It also accommodates human settlement pattern and physical results of past and present human activity (terracing, water storage or drainage structures, roads, buildings, etc.)."

If land is utilized carefully it can be considered as a renewable resource. The roots of trees and grasses bind the soil.

Land on earth is as finite as like any of our other natural resources. While mankind has learnt to adapt his lifestyle to various ecosystems world over, he cannot live comfortably for instance on polar ice caps, on under the sea or in space in the foreseeable future. Thus a comfortable land surface is important. Land is required as a resource

- Man needs land for building homes, cultivating food, maintaining\domestic animals
- developing industries to provide goods, and supporting the industry by creating towns and cities
- Equally importantly, man needs to protect wilderness area in forests, grasslands, wetlands, mountains, coasts, etc. to protect our vitally valuable biodiversity
- Man does all his activities centered around land which has a gravitational force to hold him comfortably.
- All natural resources are embedded in land for the healthy and happy life.
- Land is man's recreation center, where he aims, play and achieves his needs.

Land Degradation

Land degradation can be considered in terms of the loss of actual or potential productivity or utility. The term 'degradation' refers to irreversible decline in the 'biological potential' of the land.Every year, between 5 to 7 million hectares of land worldwide is added to the existing degraded farmland. When soil is used more intensively by farming, it is eroded more rapidly by wind and rain.

Sl.No.	Causes	Consequences	Control
1	Over irrigating farmland leads to increase in salinity	Soil loses its fertility and thus productivity decreases	Crop rotation and prevent soil erosion, water logging, etc.
2	Oil erosion removed lose water absorbing porous top soil.	Desertification starts and land becomes useless.	Not to cultivate more than three times per year.
3	Excessive chemical fertilizers used made the soil non-porous.	Soil texture is changes. Thus will support the normal life forms.	There shall be brief gap before each farming.
4	Open sewage system, industrial waste dumping.	Land loses its water carrying capacity and thus becomes water logged	Scientific method of cultivation.
5	Development of slums and their dwelling activities	Does not support for healthy life and this biodiversity is lost.	Avoid excessive use of fertilizers. Encourage organic farming practices.

Man Induced Landslides

Landslides are the downward and outward movement of a slopes composed of natural rock, soils artificial fills, or combinations thereof. Man is the cause for landslides by unscientific mining and mineral extraction activities. Constriction of dams in sensitive area causes landslides.

Characteristics of Landslides

Sl.No.	Causes	Consequences	Control
1	Natural movements of earth plates	Structural damage. Soil erosion also occurs.	Avoid deforestation and over grassing in slope areas.
2	Earthquakes	Economic and life loss	In slope area plant deep routed trees.
3	Construction of large dams	Surface geology of landslide area changed.	Avoid building huge dams in landslide prone area.
4	Mining and petroleum wells	Can not be used for construction of building since it may collapse.	Construction of concrete walls in landslide prone surfaces.
5	Water logging	Disruption of communication links, road, etc.	Contour farming, terrace farming practices, etc.

Case Study: Landslides in Kaunsa City: Natural Process Triggered by Man

The most damaging landslides were caused by irresponsible human activities. It includes excavation of slopes to enlarge the area for building or gardening. Further causes include spills from water or sewer pipes. In addition clearing of vegetation on slopes for building clearance or to have a nice view of the valley. The most evident geoindicator of land-use mistakes is more frequent landslides in the formerly stable part of the Kaunas city.

Soil Erosion

Soil erosion is the removal of top soil by man made or natural causes. Heavy wind, flood, etc constitute natural causes.

Man's activates like mining, industrialisation, etc leads to soil erosion. The roots of the trees in the forest hold the soil. Deforestation thus leads to rapid soil erosion.

Types of Soil Erosion

1. **Normal or geologic erosion**: Weathering of parent rock and erosion by agencies like water and wind are natural processes. There is always equilibrium between the removal and formation of soil, and harm is done when this equilibrium is disturbed by an outside agency.

2. **Accelerated soil erosion**: Areas denuded of their natural protective cover, excessive grazing, land that has been ploughed excessively results in the removal of top soil and thus accelerated soil erosion.

3. **Wind erosion**: Wind erosion is caused by strong wind mainly in arid and desert areas. It causes dust storms, forms sand dunes and buries localities with deposition of sand, rendering fertile lands unfit for cultivation. It is more common in Rajasthan.

4. **Water erosion**: When rain comes down in torrents there is not enough time for the water to soak through soil and the run off causes erosion.

   - Splash erosion is caused by falling torrential rain. The raindrops beat hard on the surface of the soil. The flowing mud splashes as high as 60 cm and about 150 cm away.

   - Sheet erosion occurs with the uniform removal of a thin layer or 'sheet' of soil. Sloping land with shallow, loose topsoil overlying compact subsoil is the most susceptible. This type of erosion continuously makes the soil shallower and decreases crop yield. It can be detected by the muddy colour of the run-off from the fields.

   - Rill erosion is an intermediary stage between sheet erosion and gully erosion. It involves the removal of soil by rainwater run off through small finger-like channels.

   - Channel or gully erosion occurs when the volume of concentrated run off increases and attains more velocity. The rills enlarge into gullies. It often starts along bullock cart tracks, cattle trails and burrows of animals. At an advance stage, gullies result in ravines making the soil unfit for cultivation.

5. **Landslide or slip erosion:** The pressure caused by the moisture going deep into the soil during heavy rains that is unable to penetrate further due to hard soil or rocky strata below, move a large mass of overlying soil. Such landslides are more common in hill areas.

6. **Stream bank erosion:** During a flood, the river carries large masses of soil, boulders and plants and deposits them downstream. The deposit reduces the transporting capacity of the torrent resulting in overflows and meandering of the river, and in erosion of the riverbank.

Sl.No.	Causes	Consequences	Control
1	Deforestation	Nutrient rich, productive top soil is removed thus less productive.	The use of contour plugging and wind breaks.
2	Over grassing	Lose layer of top soil which holds the crop, when lost land preparation for agriculture is costly.	Make sure that there are always plants growing on the soil, and that the soil is rich in humus (decaying plant and animal remains).
3	Hurricane, cyclone, etc.	Micronutrient requirements can not be satisfied for crops and thus leads to land become wasteland.	Avoid overgrazing and the overuse of crop lands.
4	Due to landscape topology like slope, hill, etc.	Country's agricultural product decreases and thus results in food scarcity.	Terrace farming and planting trees in slopes.
5	Flood causes soil errosion		Provide incentives to encourage farmers to manage their land sustainably.

Desertification

Desertification is a form of land degradation occurring particularly, generally in semi-arid areas leading to become desert. Increased population and livestock pressure on marginal lands has accelerated desertification. It is a misconception that droughts cause desertification.

A high population density in an area that is highly vulnerable to desertification poses a very high risk for further land degradation. Conversely, a low population density in an area where the vulnerability is also low poses, in principle, a low risk.

Sl.No.	Causes	Consequences	Control
1	Overgrazing of green area leads to desertification.	Ecological change leading to drought in such area	Well planned and scientific agricultural methods.
2	Mining and pollution can leads to desertification	People look for newer places and leading to environmental refuges.	Preserve natural ecology of a given area thereby climatic conditions are maintained.
3	Deforestation is another cause of desertification.	Loss of Biodiversity and extinction.	Construct small dams and watersheds thus better water management can be done.
4	Climate change leading to loss of monsoon can cause desertification.	The productivity of some lands declined by 50% as a result of desertification.	Land management can be performed if slope area by terrace farming, contour farming, etc.
5	Poor planning of water management in a given area can cause desertification.	It is estimated that about 500 million people affected by desertification.	Aforestation.

Role of an Individual in Conservation of Natural Resources

Education is an important tool in reducing our impact on environment. With new challenges in energy and water use, resource conservation education is playing an important role in creating individual awareness.

Individual must understand the significance of natural resources. Natural resources must be preserved and utilizing optimally.

And no natural resource is limitless. 'Non-renewable' resources will be rapidly exhausted if we continue to use them as intensively as at present.

The three most damaging factors leading to the current rapid depletion of all forms of natural resources are increasing 'consumerism' on the part of the affluent sections of society rapid population growth Greed for material goods has become a way of life for a majority of people in the developed world.

Individuals must try to understand the concepts,

- Sustainable use of our environment
- Temporary dweller on the planet Earth.
- Need to be handed over to next generation in good condition without sanction.

Energy Conservation

- Turn off lights and fans as soon as you leave the room.
- Use tube lights and energy efficient bulbs that save energy rather than bulbs. A 40- watt tube light gives as much light as a 100 watt bulb for general use.
- Keep the bulbs and tubes clean. Dust on tubes and bulbs decreases lighting levels by 20 to 30 percent.
- Switch off the television or radio or computer as soon as the program of interest is over.

- A pressure cooker can save up to 75 percent of energy required for cooking. It is also faster.
- Keeping the vessel covered with a lid during cooking, helps to cook faster, thus saving energy.

Water Conservation

- Keep the water taps tightly closed. Leaky and damaged water taps must be replaced in time.
- Use minimum water for all household application.
- Harvest rainwater and using harvesting tanks.
- Reuse water wherever possible.
- For gardening, flushing toilets etc use low quality water.
- Use water efficient toilets.

Fuels Conservation

- Keep one automobile for one family. Wherever possible combine two or more programs together. Perform pollution tests frequently.
- All fuel using methods should be checked for leaks and other losses frequently.
- While going nearby shopping over going for jacking go by walk.
- If work place is nearby combine jacking with the duty of going to office.
- Buy and use energy efficient light weight automobile.

Forest Conservation

- As for as possible use non-timber products.
- Houses and other office places should be made simple without using timber.
- Grassing, fishing, etc., must be controlled.
- All effects should be made to reclaim the lost grassland woodland etc during favourable season like during rain.

Food Conservation

(i)   Eat required amount of food. Avoid over eating followed by removing excess food by exercise.

(ii) Cook required amount of food and use it as much as possible.

(iii) Do not waste the food instead give it to the required people before getting spoiled.

(iv) Do not store large amounts of food grains in the home and also protect from damaging insects.

Equitable Use of Resources for Sustainable Lifestyles

Sustainable development is a mechanism to develop a healthy environment without damaging the natural resources. Environmental pollution is the result of population growth, industrial growth and urbanization. Environmentalists, engineers, scientists and the people interested work together to have strategies for the protection of the environment.

Present Scenario of Environment

The nature performs it duty, giving basic amenities like living space, resources, adequately. It also has a mechanism to recycle and sustain for indefinite period of time. However, the overexploitation and over utilization causes serious concern, causes ecological damage leading to un sustainability. The present trend is growing in the following direction.

- Cutting trees, using fossil fuels and emission of green house gasses.
- Various outdated technologies, which is neither economical nor sustainable.
- Disposal of various wastes without proper treatment and care.
- Poor industrial planning and industrial growth.
- Poor/wage environmental laws with slow and modified implementation.

The Need FOR Sustainable Lifestyles

The quality of human life and the quality of ecosystems on earth are indicators of the sustainable use of resources.

There are clear indicators of sustainable lifestyles in human life.

(i)   Increased longevity

(ii) An increase in knowledge

(iii) An enhancement of income.

These three together are known as the 'Human development index'.

Sustainable development also includes

- A stabilized population.
- The long term conservation of biodiversity.
- The careful long-term use of natural resources.
- The prevention of degradation and pollution of the environment.

Equitable Distribution

A new economic order at the global and at national levels must be based on the ability to distribute benefits of natural resources by sharing them more equally among the countries as well as among communities within countries such as our own. It is at the local level where people subsist by the sale of locally collected resources, that the disparity is greatest. 'Development' has not reached them and they are often unjustly accused of 'exploiting' natural resources.

This primarily comes from caring for our Mother Earth in all respects. A love and respect for Nature is the greatest sentiment that helps bring about a feeling for looking at how we use natural resources in a new and sensitive way. Think of the beauty of a wilderness, a natural forest in all its magnificence, the expanse of a green grassland, the clean water of a lake that supports so much life, the crystal clear water of a hill stream, or the magnificent power of the oceans, and we cannot help but support the conservation of nature's wealth. If we respect this we cannot commit acts that will deplete our life supporting systems.

Thus those who have it share it, propagate it and protect it. Natural resources are common to all.

Questions

1. What is the need for environmental science for the present generation?
2. Explain forest over-exploitation with a suitable example.
3. Substantiate the statement "depletion of water resources is due to lake of sustainable approach".
4. What are environmental effects due to extraction of minerals?
5. Food is a serious concern to the population-explain.
6. Outline the approaches to solve the food demand and its environmental consequences.
7. Global energy consumption is increasing-solve the problem by methods of renewable energy concept.
8. Distinguish renewable and non-renewable energies.

9. Land degradations are as a consequence of over utilization of land resources-explain.

10. What are the strategies adopted to conserve natural resources at individual level?

11. What is meant by sustainable lifestyle? How will you achieve it?

Social Issues and the Environment

From unsustainable to sustainable development – urban problems related to energy – water conservation, rain water harvesting, watershed management – resettlement and rehabilitation of people; its problems and concerns, case studies – role of non-governmental organization- environmental ethics: Issues and possible solutions – climate change, global warming, acid rain, ozone layer depletion, nuclear accidents and holocaust, case studies. – wasteland reclamation – consumerism and waste products – environment production act – Air (Prevention and Control of Pollution) act – Water (Prevention and control of Pollution) act – Wildlife protection act – Forest conservation act – enforcement machinery involved in environmental legislation- central and state pollution control boards- Public awareness.

From Unsustainable to Sustainable Development

According to Norwegian Prime Minister and Director of World Health Organization, G.H. Bundland "Sustainable development is meeting the needs of the present without compromising the ability of future generations to meet their own needs.

The discussion on sustainable development was held in 1992 UN Conference on Environment and Development (UNCED) also known as Earth Submit, held Rio de Janeiro, Brazil. The popularly known Rio Declaration states that, " a new and equitable global partnership through the creation of new levels of cooperation among states...".

1. **Reduce, Reuse and Recycle (3R) Approach:** The aim of the approach is minimum consumption, use it further till it gives its maximum utility and then recycle it into further product instead of dumping as waste. The concept is widely accepted and many countries including India is practicing it.

2. **Technological Improvement:** In this direction many efforts have been made to increase the efficiency of existing technology by improving techniques for waste reduction and promote maximium production from a given input. Further thrust for new technologies has also been motivated. Presently by considering plastic age biodegradable plastics are manufactured by few companies.

3. **Consumption of Non-renewable Resources:** Stringent norms have been imposed on the use of Non-renewable resources. Alternatives for such resources have also been invoked.

4. **Zero Waste Concept:** Convert the waste generated into wealth and such that nothing is left out waste. In other words everything is used for some purpose or other in order to reduce the waste to zero.

5. **Sustainable Lifestyles:** Adopt and live in such a way nothing in the environment is destroyed for the personal comfort. Forgo some part of comfortable life for the purpose of environmental sustainability.

6. **Use Twelve Principles of Green Chemistry.**

    1. **Prevent waste:** Design chemical syntheses to prevent waste, leaving no waste to treat or clean up.

    2. **Design safer chemicals and products:** Design chemical products to be fully effective, yet have little or no toxicity.

    3. **Design less hazardous chemical syntheses:** Design syntheses to use and generate substances with little or no toxicity to humans and the environment.

    4. **Use renewable feedstocks:** Use raw materials and feedstocks that are renewable rather than depleting. Renewable feedstocks are often made from agricultural products or are the wastes of other processes; depleting feedstocks are made from fossil fuels (petroleum, natural gas, or coal) or are mined.

    5. **Use catalysts, not stoichiometric reagents:** Minimize waste by using catalytic reactions. Catalysts are used in small amounts and can carry out a single reaction many times. They are preferable to stoichiometric reagents, which are used in excess and work only once.

    6. **Avoid chemical derivatives:** Avoid using blocking or protecting groups or any temporary modifications if possible. Derivatives use additional reagents and generate waste.

    7. **Maximize atom economy:** Design syntheses so that the final product contains the maximum proportion of the starting materials. There should be few, if any, wasted atoms.

    8. **Use safer solvents and reaction conditions:** Avoid using solvents, separation agents, or other auxiliary chemicals. If these chemicals are necessary, use innocuous chemicals.

    9. **Increase energy efficiency:** Run chemical reactions at ambient temperature and pressure whenever possible.

    10. **Design chemicals and products to degrade after use:** Design chemical products to break down to innocuous substances after use so that they do not accumulate in the environment.

11. **Analyze in real time to prevent pollution:** Include in-process real-time monitoring and control during syntheses to minimize or eliminate the formation of byproducts.

12. **Minimize the potential for accidents:** Design chemicals and their forms (solid, liquid, or gas) to minimize the potential for chemical accidents including explosions, fires, and releases to the environment.

Urban Problems Related to Energy

Human development is associated with 'urbanization' and is inevitable. This leads to increase in population density in urban areas and hence energy needs also in increase. Indeed the per capita energy consumption also high in urban areas due lifestyles, entertainment activities and other cultural and social activities.

Sl.No.	Energy Requirements	Energy Problems	Solution
1	Residential and commercial lighting	Shortage of energy supply	Minimization of energy use
2	Public and private transportation	Decrease the availability like power cut	
3	Electrical and electronic appliances	Black market sale	Energy audit
4	Industrial requirements	Adulteration like mixing kerosene.	Increase production capacity
5	Requirements for recycle, waste treatment, etc	Energy theft	Use of energy efficient technology
6			Imposing strict laws and penalties.

Water Conservation

- Though water is available plenty but not in the required way and purity or not where and when it is required. Especially good quality water is always a scarce commodity and constantly invoked ways and methods to get it and optimally use it. The principal source of fresh water is rain and is considered to be the purest in ideal environmental condition. In addition water is available in rivers, streams, lakes, ponds, and also plenty in the sea. In every year 22 March is observed as World Day.

Need For Conservation

- As the population increases the water requirement also more.
- Though resources are plenty the quality and reliability are not high due to variation in environmental factors.
- Further due to deforestation etc the annual availability of water is in constant decrease.
- Industrialization enhance the industrial demand of water and many times it requires fairly good quality a also in large volumes.

* Better lifestyles demand more water.

Water Conservation Strategies

* Reduce the usage of water. Reuse and recycle the water as much as possible
* Invoke water efficient technology and also house hold activity like detergents with less leather and better cleansing activity.
* Conserve water when it is available in excess. Water harvesting and watershed management strategies.
* Preserve the natural water by methods like rainwater harvesting, building dams, saving in ponds, lakes etc.
* Not to pollute the water by sewage, effluents, radiations, etc.

Rain Water Harvesting

It is a method adopted to capture and store rain water for further utilization. As the year passes the precipitation pattern is unpredictable and decreasing. This leads to the evolution of concept called rainwater harvesting. Since the rainwater is in general pure water and its value is high thus can not afford to loose it.

Rainwater harvesting is the principles and methods, which avoids loses of rainwater by collecting and preserving for future use. Villages traditionally use the methods like storing rainwater in community reservoirs like ponds, lakes etc.

Importance of Rainwater Harvesting

* Rainwater harvesting also decreases the chance of local floods consequently damage and distraction to public person and property.
* It increases the chance of water availability when it is required and avoids water scarcity.
* The harvesting methods also increase the underground water levels and can be utilized when it is required using appropriate technology.
* Prevents ground water contamination in coastal area.
* Reduce run off loss and soil erosion.

Harvesting System

Broadly rainwater can be harvested for two purposes.

* Storing rainwater for ready use in containers above or below ground.
* Charged into the soil for withdrawal later (groundwater recharging).

Rainwater harvesting can be harvested from the following surfaces

Rooftops

If buildings with impervious roofs are already in place, the catchment area is effectively available free of charge and they provide a supply at the point of consumption.

Paved and Unpaved Areas

Landscapes, open fields, parks, stormwater drains, roads and pavements and other open areas can be effectively used to harvest the runoff. The main advantage in using ground as collecting surface is that water can be collected from a larger area. This is particularly advantageous in areas of low rainfall.

Water Bodies

The potential of lakes, tanks and ponds to store rainwater is immense. The harvested rainwater can not only be used to meet water requirements of the city, it also recharges groundwater aquifers.

Components of a Rainwater Harvesting System

A rainwater harvesting system comprises components of various stages - transporting rainwater through A roof made of reinforced cement concrete (RCC), galvanised iron or corrugated sheets can also be used for water harvesting.

1. **Coarse mesh:** It prevent the passage of debris
2. **Gutters:** Channels all around the edge of a sloping roof to collect and transport rainwater to the storage tank. Gutters can be semi-circular or rectangular and could be made using:
   - Plain galvanised iron sheet (20 to 22 gauge), folded to required shapes.
   - Semi-circular gutters of PVC material can be readily prepared by cutting those pipes into two equal semi-circular channels.
   - Bamboo or betel trunks cut vertically in half.

   The size of the gutter should be according to the flow during the highest intensity rain. It is advisable to make them 10 to 15 per cent oversize.
3. **Conduits:** Conduits are pipelines or drains that carry rainwater from the catchment or rooftop area to the harvesting system. Conduits can be of any material like polyvinyl chloride (PVC) or galvanized iron (GI), materials that are commonly available.

4. **First-flushing:** A first flush device is a valve that ensures that runoff from the first spell of rain is flushed out and does not enter the system. This needs to be done since the first spell of rain carries a relatively larger amount of pollutants from the air and catchment surface.

5. **Filter**

   The filter is used to remove suspended pollutants from rainwater collected over roof. A filter unit is a chamber filled with filtering media such as fibre, coarse sand and gravel layers to remove debris and dirt from water before it enters the storage tank or recharge structure. Charcoal can be added for additional filtration.

6. **Sump:** A storage provision is used to collect filtered water. It is from the tank appropriate size and kept in safe place. The water is available for immediate use.

7. **Soakaways / Percolation pit:** Percolation pits, one of the easiest and most effective means of harvesting rainwater, are generally not more than 60 x 60 x 60 cm pits, (designed on the basis of expected runoff as described for settlement tanks), filled with pebbles or brick jelly and river sand, covered with perforated concrete slabs wherever necessary. This is used for ground water recharge.

Watershed Management

Watershed is defined as a geohydrological unit draining to a common point by a system of drains. Watershed is thus the land and water area, which contributes runoff to a common point. "Watershed management" is the principles and practices followed for effective use and utilization of watershed. It is a village level activity.

Watershed management involves the judicious use of natural resource with active participation of institutions, organizations, in harmony with the ecosystem.

Importance of Watershed Management

1. It is water storage system. Thus decreases the local flood, soil erosion, water logging, etc.
2. Minimize water scarcity and draught, desertification, etc.
3. It makes the community to use water when seasonal rain fails.
4. Promotes agriculture in the surrounding area.
5. Preserve ecosystem and biodiversity in local level.
6. Water loss is prevented during rainy season.
7. It promotes rural life style and self sufficiency.
8. It promotes employment opportunities, livelihood and security.

9. It avoids ground water contamination and increase the ground water level.

10. It gives permanent solution to "water prosperity".

Types of Watershed

The common modes of categorization are the size, drainage, shape and land use pattern. The categorization could also based on the size of the stream or river, the point of interception of the stream or the river and

1. Macro watershed (> 50,000 Hect).
2. Sub-watershed (10,000 to 50,000 Hect).
3. Milli-watershed (1000 to10000 Hect).
4. Micro watershed (100 to 1000 Hect).
5. Mini watershed (1-100 Hect).

Watersheds may also be categorized as hill or flat watersheds, humid or arid watersheds, red soil watershed or black soil watershed based on criteria like soil, slope, climate etc.

Depending on the land use pattern watershed could again be classified as highland watersheds, tribal settlements and watersheds in areas of settled cultivation.

Watershed Management at the Village Level

The Indo-German Watershed Program started in 1992. It is managed by the National Bank for Agriculture and Rural Development (NABARD) and Watershed Organization Trust (WOTR) in Ahmednagar.

Germany provides grant finance and some technical and administrative back-stopping. Coverage of the program has increased from 22 villages in 1995 up to 83 in November 1997. The villages are in various regions of Maharashtra. Today, about 100.000 people on 244.000 acres of land have directly profited from the program.

This system makes it necessary for the villages to organize themselves right from the beginning of the cooperation with the program.

During recent years, many Village Watershed Committees, self-help groups and women's groups have emerged, with the encouragement and training from WOTR and other NGOs.

Watershed Management Techniques

It is the methods used to construct catchment area for water storage.

Trenches (Pits): These are pits made on ground. It accommodates small water and helps in increase the ground water level.

Earthen Dam or Stone Embankment: It prevents surface run off and allows to stay the water in the catchment area.

Farm Pond: It is a water storage area with definite boundaries. The water run off is collected in this area during rainy season. The water is mainly used for agriculture.

Underground Barriers (Dykes): It is a barrier build below ground level when the running water body is trough deep pit. The barrier allows the water to be stagnant and percolate into the surround ground thus underground water level gets increased.

Components of Watershed Management

The three main components in watershed management are land management, water management and biomass management.

Land Management

Land characteristics like terrain, slope, formation, depth, texture, moisture, infiltration rate and soil capability are the major determinants of land management activities in a watershed. The broad category of land management interventions can be as follows;

- Structural Measures.
- Vegetative Measures.
- Production Measures.
- Protection Measures.

Structural measure include interventions like contour bunds, stone bunds, earthen bunds, graded bunds, compartmental bunds, contour terrace walls, contour trenches, bench terracing, broad based terraces, centripetal terraces, field bunds, channel walls, stream bank stabilization, check dams etc.

Watersheds may contain natural ecosystems like grasslands, wetlands, mangroves, marshes, water bodies.

All these ecosystems have a specific role in nature. Vegetative measures include vegetative cover, plant cover, mulching, vegetative hedges, grassland management, vettiver fencing, agro-forestry, etc.

The production measures include interventions aimed at increasing the productivity of land like mixed cropping, strip cropping, cover cropping, crop rotations, cultivation of shrubs and herbs, contour cultivation conservation tillage, land leveling, use of improved verity of seeds, horticulture, etc.

Protective measures like landslide control, gully plugging, runoff collection, etc can also be adopted. Adoption of all the interventions mentioned above should be done strictly in accordance with the characteristics of the land taken for management.

Water Management

Water characteristics like inflows (precipitation, surface water inflow, ground water inflow) water use (evaporation, evapotrasnpiration, irrigation, drinking water) outflows (surface water outflow, ground water out flow) storage (surface storage, ground water storage, root zone storage) are the principal factors to be taken care of in sustainable water management.

Biomass Management

Major intervention areas for biomass management are indicated below;

- Eco-preservation
- Biomass Regeneration
- Forest Management & Conservation
- Plant Protection & Social Forestry
- Increased Productivity of Animals
- Income & Employment Generation

People's Participation

People's participation and collective action are critical ingredients for watershed management. Sustainability, equity and participation are the three basic elements of participatory watershed management.

This includes natural resource exchange, which is the conventional watershed management, and participatory watershed management additionally considers the economic, political and cultural exchanges.

But the advantages of people's participation are many and sound. Participation can ensure effective utilization of available resources. In real terms community participation means voluntary sharing by the users group their time, energy and money on the program and adopt the recommended measures and practices on a sustained basis.

Resettlement and Rehabilitation of People: Problems and Concerns

Resettlement is displacement of people in newer area. Rehabilitation is the process of making the displaced people comfortable in the newer habitat.

Need for Resettlement

- Government policies like industrial zones, special economic zone formation.
- Natural calamities and make the place dangerous for further living.
- Building dams and related activity like power project.
- Research area, army occupation, etc.

- Nuclear power projects.
- Abolition of slums and urbanization, etc.

Need for Rehabilitation

- The population can not live comfortably in the new environment.
- All basic amenities has to be provided.
- Basic requirement for life sustainability have to be ensured.
- Public health clinics, schools, transport, lighting, communication link, etc has to be established.
- Right to dignified life is fundamental right as per Constitution of India. It has to be ensured by the Government of India to its citizens with out expectation.

Problems (Issues)

- Tribal people unwilling to loss their natural habitat.
- Poor citizens depending on natural resources are put into unsustainable area.
- Community life is affected and social disintegration occurs.
- People losses the property and values in their original dwelling place.
- Land acquisition laws and displacement policies are unimaginable to these paupers.
- Giving equivalent property, adequate compensation etc are only in paper and records.
- People in the new environment start undergo adaptation. This leads to manufacture of criminals, sex traders, illegal business, etc. initially for survival.
- Participation of women workers and child labourers in Indian economy.

Concerns

- Money lenders, terrorist organizations explore their business and subject them to overexploitation.
- Many Government authorities become rich by corruptive practices on rehabilitation account.
- Conflict between people and police in coordination with non-government and social organizations.
- Begging to World Bank in terms of rehabilitation currency.
- Poor facility and care leading to hate and crime increase.
- Conflict between local people and displaced population.
- Difficulty in getting titles in the new place, identity and establishment of recognition.
- Traditional and family profession can not be practiced.

Rehabilitation Policy

The draft bill of The Rehabilitation And Resettlement Bill, 2007 has been made make the rehabilitation, effective and ethical.

- In case the entire population of the village or area to be shifted belongs to a particular community. In the case of resettlement of the Scheduled Castes affected families, such families may, wherever possible, be resettled in the areas close to the villages. (Section 26).
- If house is lost area for building house to be give as per section 35 of the bill.
- If land lost given equivalent land as per Section 36.
- If cattle shed is lost minimum 15000 rupees to be paid (Section 38)
- Atleast rupees 10000 to be paid for shifting the family (Section39)
- A minimum of 25000 rupees for artisans for building working shed or shop (Section 40)
- Employment to affected family in the project (Section 41).
- Agricultural income loss is substituted by providing 150 days of agricultural income (Section 42)
- For vulnerable persons in the family to get Rs 500 pension (section 46)
- Tribal Development plan (Section 49).

As a whole many more benefits derived for displaced people (as per the Bill). If such procedures are followed there will be minimal impact due to displacement of people.

Case Studies

Narmada Water Disputes Tribunal (NWDT) Award

Narmada Dam or Sardor Sarovar Dam Project is construction of Big and smaller dams across Narmada River in Maharastra. Initially World Bank agreed to finance the project and later withdrew from it.

Due to Writ Petition (Civil) 319 of 1994 of Narmada Bachao Aandolan (NBA) vs. Union of India and others.

A mechanism has been evolved to link the progress of R&R works and dam construction keeping in view the interim orders/ directions of the Hon'ble Supreme Court. Accordingly, the programs of dam construction will be approved on year to year basis after reviewing the progress of R&R works.

Out of total of 40,827 project affected families (PAFs), house plots have been allotted to 10,471 families and agricultural land has been allotted to 9,998 families upto 31st January, 2000.

The project affected families are also provided subsistence allowance, rehabilitation grant, ex-gratia, productive assets, insurance cover and civic amenities like primary schools, dispensaries, children park, panchayat ghar, religious places, tree platforms, wells, hand-pumps, transit sheds, electrification, etc. and employment to some of them.

However, many reports and documents are bogus. Indeed social activist Ms. Medha Patkar performed indefinite hunger strike in this regard and she called off because of the request of President Dr. APJ Abdul Kalam.

Role of Non-Government Organizations (NGOs)

Non-Government organizations are organized group of people who have similar interest, vision and mission in relation to welfare of people, property and environment. They can be registered entitities can enjoy the rights and liabilities like unions. Some of the roles of are described below.

(i) Awareness Programs

- NGO's can propagate knowledge to general public regarding many issues like environment, sustainable life style, family welfare programs, etc. Also brought to the knowledge of Govt.

(ii) Rights and Liabilities

- They can make the public in rights and duties to be performed to the nations. They can demonstrate and assist in getting eligible share of rights when they are displaced or relocated, etc.

(iii) Peaceful Demonstration

- They can coordinate and call for public bundh and other peaceful demonstration mechanism in order to get the eligible remedy for the people.

(iv) Public Interest Litigation

- NGO's can file Public Interest Litigation in High Courts and Supreme Court of India in maters concerning to the interest of larger public.

(v) Right to Representation

- Where there is necessity NGO's can represent on behalf of public or person and derive benefits to the affected person.

(vi) Right to Legal Remedy

- NGO's are welcome to file legal petitions with appropriate authorities or in Courts on behalf of another person or public. They can file Writ petitions and represent the same.

Examples of NGO's (National)

a) **CARPED:** The main objective of the organization is to facilitate better quality of life in all its realms through community mobilization, participatory governance based on sustainable natural resource management.

b) **Anakkara Vikasana Sangam:** AVS aims to promote social, cultural, economic and scientific development and environmental protection of the weaker sections and economically backward people of society.

c) **Child Rights and You (CRY):** Looking after the child development, education, abolition of child labour, Human Right relevant to children, etc.

d) **Environmental Society of India-Chandigarh**

e) **Madras Environmental Society-Chennai, etc.**

International NGO's

1. Center for International Law.
2. Conservation International.
3. International Forest Policy.
4. The World Conservation Union, etc.

Environmental Ethics

Introduction

As long as man was contended with what he had, when he had, when he was not subdued by greed he lived a happy life. But when he wanted to control nature, when he wanted to reign supreme in this universe, when pampering of the flesh and starvation of the soul began he started defining his environment. Now after enough medaling with environment started about environmental ethics. Ethics is moral principles. Morality is socially acceptable norms and practices. Environmental ethics is to be moral to the environment.

Ethical Practices

- Sustainable life style.
- Minimum utilization maximum comfort.
- 3R policy.

- Love peace and solve disagreements with dialogues in stead of war.
- Ban all nuclear arsenals and research in such direction.
- Respect nature and its creations.
- Respect all living forms including inanimate things on the planet.

Conclusion

Wordsworth cautioned us "Our meddling intellect, mishaps the beauteous forms of things, we murder to dissect". When he saw before him men and women falling dead like flies before a burner when the 'Litle Boy' and 'Fat Man' were dropped on Nagasaki and Hiroshima and the Chronobyl and Bhopal tragedies, he realized his folly and wanted to make amends.

The pesticides, insecticides and fertilizers used to enrich yield and tests of some materials cause air pollution leading to diseases such as cancer. In 1958 Paul Muller was awarded Noble prize for his discovery of DDT. Though malaria was eradicated, it is banned due to cumulative toxin.

Man realized that he would kill himself if he thoughtlessly and violently upsets the delicate environment or which he is a part. Thus we need to practice environmental ethics.

Environmental Impact Assessment (EIA)

Environment Impact Assessment or EIA can be defined as the study to predict the effect of a proposed activity/project on the environment.

A decision making tool, EIA compares various alternatives for a project and seeks to identify the one which represents the best combination of economic and environmental costs and benefits.

Three Core Values of EIA

Integrity: The EIA process should be fair, objective, unbiased and balanced

Utility: The EIA process should provide balanced, credible information for decision-making

Sustainability: The EIA process should result in environmental safeguards

EIA systematically examines both beneficial and adverse consequences of the project and ensures that these effects are taken into account during project design. It helps to identify possible environmental effects of the proposed project, proposes measures to mitigate adverse effects and predicts whether there will be significant adverse environmental effects, even after the mitigation is implemented.

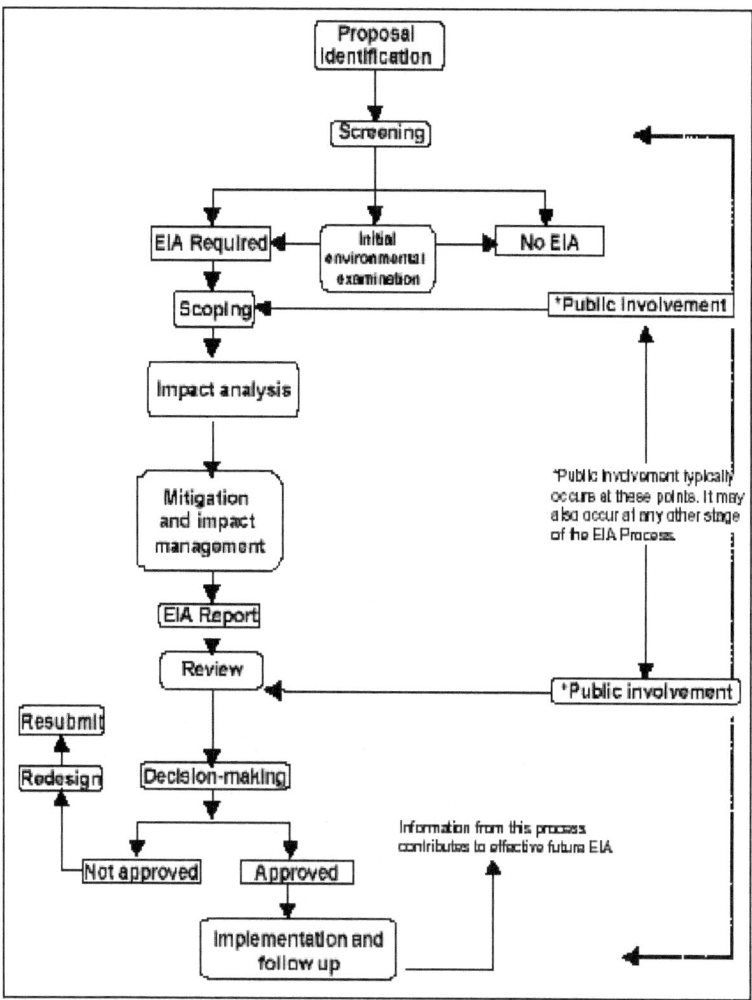

The EIA Process

Main stage is a basic standard of good practice followed in EIA process. Though, it varies from project to project and country to country.

The environment impact assessment consists of eight steps with each step equally important in determining the overall performance of the project.

1. **Screening:** First stage of EIA, which determines whether the proposed project, requires an EIA and if it does, then the level of assessment required.

2. **Scoping:** This stage identifies the key issues and impacts that should be further investigated. This stage also defines the boundary and time limit of the study.

3. **Impact analysis:** This stage of EIA identifies and predicts the likely environmental and social impact of the proposed project and evaluates the significance.

4. **Mitigation:** This step in EIA recommends the actions to reduce and avoid the potential adverse environmental consequences of development activities.

5. **Reporting:** This stage presents the result of EIA in a form of a report to the decision-making body and other interested parties.

6. **Review of EIA:** It examines the adequacy and effectiveness of the EIA report and provides the information necessary for decision-making.

7. **Decision-making:** It decides whether the project is rejected, approved or needs further change.

8. **Post monitoring:** This stage comes into play once the project is commissioned. It checks to ensure that the impacts of the project do not exceed the legal standards and implementation of the mitigation measures are in the manner as described in the EIA report.

Participants of EIA Process

Agency	Deacription
Proposer	Individual or organization
Decision Making Authority	Ministry or Head of State
Assessor	Responsible EIS person
Reviewing Agency	Agency or Individual with expertise
Experts	Persons with special knowledge and skills
Public Member/Community	Citizens who has interest or affected by the project.
Mdia	Print and electronic media
Special Interest Groups	Government and NGO's Unions etc.

Benefit of EIA

1. A controlled scientific plan of course of action.
2. Public participation and reliable conclusion.
3. Alternatives are considered and best being selected.
4. Project feasibility is in mind, if not dropped.
5. Cost benefit analysis also performed.
6. A healthy sustainable development practice.

Environmental Issues and Possible Solutions

Climate Change

It is the observed phenomenon of changes in the weather pattern of a particular area over a period. It includes seasonal variations, day and night temperatures, precipitation pattern, window flow and directions, etc.

Sl.No.	Causes	Consequences	Control
1	Changes in the composition of ambient air.	Global warming and increased sea level.	Follow emission norms.
2	Excessive human activity (urbanization)	Changes in ecosystem and its diversity.	Treat the effluent and industrial emissions
3	Industrialization, Emission of CFC's	Changes in ocean current, wind flow directions, precipitation patterns, etc.	Alternate ecofriendly technologies
4	Deforestation	Migration of birds and animals.	Aforestation
5	Changes in agricultural pattern.	Food scarcity, disrupted food chains, etc.	Follow Green Chemistry principles in industry

Global Warming

It is the process of increase in temperature of earth's atmosphere than the expected rate. It is proposed that the global warming is due to the accumulation of gasses termed as green house gases in the environment. Carbon dioxide, chlorinated hydrocarbons (CFS's), methane, etc. are considered as green house gases.

Solar radiations are heat radiations. After sticking the earth's surface re-radiate to space at higher wavelength. The CO_2 and green house gases absorb longer wavelength radiations and get heated up. They radiate heat energy to earth surface and warms up. This is the origin of global warming.

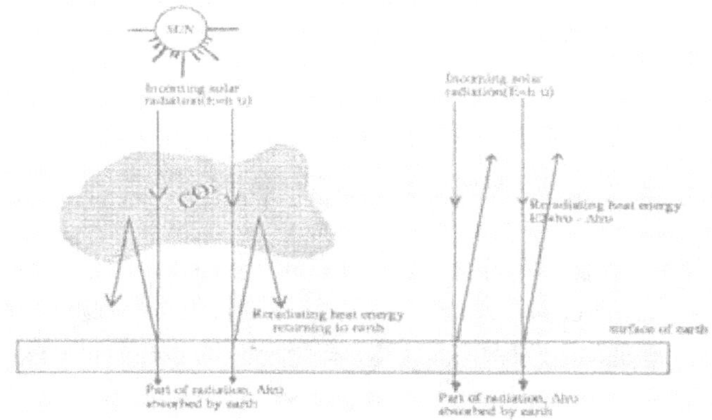

Sl.No.	Causes	Consequences	Control
1	Industrial Emissions of Green House gases	Climate change	Decrease the CFCs to sustainable limit.
2	Deforestation leading to decreased utilization of CO_2	Increase in sea level and displacement of people.	Control industrial emissions
3	Ozone hole leading to increased penetration of UV rays.	Changes in ecosystem and biodiversity.	Strict emission standards for automobiles.
4	Use of refrigeration system with CFC's	Decrease in agricultural productivity and food scarcity.	Aforestation
5		Spread of disease, affects human health.	Use of green energy concepts.

Trichloro fluoromethane (CCl_3F), Dichlorodifluoromethane (CCl_2F_2), Chlorotrifluromethane (CCF_3) are some of the commonly used CFC's in refrigeration engineering and manufacture of foams, polymers, etc.

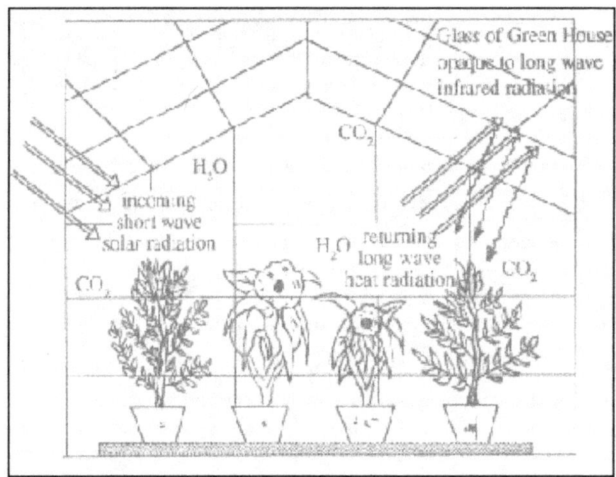

Green House Effect (GHE)

A glass experimental containing CO_2, water vapour allowed to keep the container hot and helped the green vegetation to grow well. The gas is called as green house gas. Later other gases like chloroflurocarbons, methane, oxides of nitrogen and sulfur, etc also added as green house gases.

Glass container is transparent to shorter wavelength solar radiations. They are reflected to longer wavelength. The CO_2 and Water vapour aborbed the longer wavelength radiations and gets heated up. The heated molecules re-radiate heat energy and maintained the temperature of the chamber. Thus the optimal temperature provided ideal environment for the growth of green plants. Thus they got the award green house gases.

Sl.No.	Causes	Consequences	Control
1	Industrial emissions	Increased sea level, due to water expansion and melting of ice.	Control the emission of green house gases.
2	Deforestation	Climate change	Aforestation
3	Use of fossil fuels	Global warming, Ozone layer depletion.	Improve automobile emission standards
4	Automobile emissions	Health problems, spread of disease	Avoid the unsustainable use of fossil fuels
5	Green Revolution	Changes in agriculture	Organic Farming

Ozone Layer Depletion

Earth has a protective umbrella in the form of Ozone Layer of 24 km thickness in the stratosphere about 15 km away from the earth's surface. The concentration of ozone in stratosphere is about 10 ppm. It is essential for the life to sustain on earth. It absorbs the dangerous ultraviolet radiation (UV-γ\rays with wavelengths from 200-280 nm) from the sun and converts it to heat and chemical energy.

In nature, ozone is continuously formed and destroyed through photochemical interaction and equilibrium in ozone concentration is ensured. However this equilibrium is upset due to air pollutants such as Chloro Fluoro Carbons (CFSs) into the atmosphere. The CFCs release free radicals of chlorine, fluorine and bromine. It destroy the stratospheric ozone. As a result of which the ozone layer is thinned. The patches of thinned ozone layers are known as "Ozone Holes".

Mechanism of Ozone Depletion

CFCs absorbs UV radiations present in the stratosphere and decompose them into chlorine and fluorine radicals.

$$CF_2Cl_2 \xrightarrow{\text{sun light}} CF_2\overset{\bullet}{Cl} + \overset{\bullet}{Cl}$$
$$\text{Freon-12} \qquad\qquad\qquad \text{Chlorine radical}$$

$$CF_2Cl \xrightarrow{\text{sun light}} \overset{\bullet\bullet}{CFCl} + \overset{\bullet}{F}$$
$$\text{Fluorine radical}$$

Then free radicals react with ozone, disintegrate it into oxygen and oxygen radical It is estimated that each atom of chlorine radical can destroy 1,00,000 molecules of ozone catalytically, as shown below.

$$\overset{\bullet}{Cl} + \overset{\bullet}{O_3} \longrightarrow \overset{\bullet}{ClO} + \overset{\bullet}{O_2}$$
$$\text{Ozone}$$

$$O_3 \longrightarrow O_2 + \overset{\bullet}{O}$$
$$\text{Oxygen radical}$$

$$\overset{\bullet}{ClO} + \overset{\bullet}{O} \longrightarrow \overset{\bullet}{Cl} + O_2$$
$$\text{Propagating radical}$$

The Montreal Protocol on Substances That Deplete the Ozone Layer (a protocol to the Vienna Convention for the Protection of the Ozone Layer) is an international treaty designed to protect the ozone layer. The treaty was opened for signature on September 16, 1987 and entered into force on January 1, 1989 followed by a first meeting in Helsinki, May 1989. India respectfully ratified the treaty.

Sl.No.	Causes	Consequences	Control
1	Industrial emission of CO_2.	Ozone layer depletion	Emission control norms
2	Emission of CFC's from refrigeration	Global warming	Altharnative technology for CFC's
3	Use of Fossil fuels	Heath problems, skin diseases, birth defects	Automobile emission standards.
4	Automobile emissions	Climate change	Avoid using fossil fuels
5	Urbanization	Ecological imbalance	Aforestation

Acid Rain

The pH of rain water is in the moderate to strong acid is known as acid rain. Acid rain in general has the pH less than 5.6. In developed countries acid rains with pH <4.5 are common.

Thermal power plants, industries and other sources release thousand of tones of oxides of nitrogen and sulphur into the atmosphere everyday. These gases undergo transformation in the atmosphere form nitrates, sulphates, notric acid or sulphuric acid droplets. Some of these pollutants can travel 200-300 km in a day. Thus compound emitted in one place may cause effect of concern on another place.

Rain is the purest source of water. 'Acid Rain' means any precipitation-rain snow, or dew, which is more acidic than normal.

It is estimated that about 70% of acidity of acid rain is due to SOX (Sulfur Oxides) and 30% due to NOX (Nitrogen Oxides) emissions. Majority of fuels, especially coal contain 0.5-4% and when burnt in air huge quantities of SOX and NOX are released. The super stack and the mammoth smelters at Sudbury of Ontario release about 2500 tons of SO_2 everyday account about 1% of global contribution.

Sl.No.	Causes	Consequences	Control
1	Industrial emission of oxides of nitrogen and sulfur	Algae population in water bodies depleted	Control of emission of nitrogen and sulphur oxides.
2	Automobile emissions	Corrosion of structures and machines	Automobile emission standards.
3	Use of fossil fuels.	Heath problems and diseases	Use of green vehicles for transport
4	War between countries	Water contamination.	Green energy concepts.
5	Chemical pollution	Lung and heart problems.	Green Chemistry principles.

In 1958, the rain falling over Europe had a pH of 5.0. It was 4.5 in 1962 in Netherlands ands Sweden. In 1970, ecological damage was observed while in 1979, it was observed that 20,000 lakes of Sweden alone were suffering grom loss of flora and fauna. In USA, in 1979, the average pH value of rain was 4.2. IN Los Angles, a pH of 3.0 was notice for a dense fog. Wheeling (West Virgina) of USA and Pitlochry of Scotland recorded pH of about 2.2, is as acid as battery acid. The rain on April 10, 1984 at Pitlochry was nearly one lakh times more acid than normal water (less than pH 2).

Black Snow

During the Gulf war of 1991, thousands of tones of smoke and dust were released due to burning of hundreds of oil wells.

A portion of these pollutants carried by wind were finally deposited on the shining skin of the Himalayas as a "Black Snow". Hundreds of acres of the Himalayas were covered by a black carpet of smoke and soot. This would not only damage the beauty of the Himalayas but also would lead to higher melting of snow due to the higher absorption of solar radiation by the black surface. The flora and fauna of the mountain also may be greatly affected when exposed to such pollutants.

Nuclear Accidents and Holocaust

Holocaust is a large scale murdering of human being by human beings. Historically Mr. Hitlor performed such operations on Jews population thereby initiated II World War. As nuclear energy release is enormous and thus holocaust becomes much easier as demonstrated on Japan (Nagasaki and Hiroshima) at end of II World War. Recent time nuclear accident can perform the same task.

As nuclear war heads are at present readily available. Thus further nuclear holocaust can be expected. On using so environment and ecosystem also damaged. For generations human beings with birth defects, premature birth and so on.

In addition nuclear winter is also an observable phenomenon. It is due to the accumulation of dust and black particles on the upper surface of the earth. It avoids the penetration of light into the earth surface. So the surface becomes cooler and life forms struggle to survive.

Einstein warned, "The splitting of the atom has changed everything save our modes of thinking, and thus we drift toward unparalleled catastrophe."

It will remembered and mourned the dreadful loss of innocent civilian life when the nuclear bomb - 'Little Boy' exploded 600 metres above the city of Hiroshima at 8:15am killing 140,000 people. Three days later, at 11.02am 'Fat Man' was dropped on Nagasaki resulting in the immediate deaths of 40,000 people. The spectre of a nuclear holocaust is today again a reality, not known since the Cuban Missile Crisis in 1962. The possibility of a nuclear disaster occurring has escalated following the War in Iraq.

More than a decade after the end of the Cold War, there are eight countries that possess nuclear weapons. These are the United States, Russia, United Kingdom, France, China, India, Pakistan and Israel. More than 95 percent of the world's more than 30,000 nuclear weapons are in the arsenals of the US and Russia. Some 4,500 of these remain on hair-trigger alert, ready to be fired upon a moment's notice.

This is the greatest challenge of our time, and it is a shared responsibility of all of us alive today. We hope that you will play an active part in ending the nuclear weapons era and making the world safe for all forms of life, including ourselves.

"I know not with what weapons World War III will be fought, but World War IV will be fought with sticks and stones."

-- Albert Einstein

Wasteland Reclamation

It is the process and methods for the conversion of wasteland into land suitable for use of habitation or cultivation. Approximately 20% of the geographical area of India is now under the wasteland.

Due to increased need of land surface for agricultural and other development activities wasteland is converted into productive land by scientific methods.

Formation of Wasteland

- Over exploitation
- Desertification
- Soil erosion
- Draought
- Unethical agricultural practices
- Overgrazing, forest degradation.
- Mining and oil extraction processes
- Pollution and dumping of solid and liquid wastes.

Reclamation Practices

1. Test the soil and analyze the nutrient requirements and perform appropriate treatment to make it cultivable land.
2. Perform biological treatment if the soil is highly acidic or basic.

3. Select crop varieties appropriate to the soil without the addition of large amount of manures.

4. Green manure and bio fertilizers to be used to in preference to chemical fertilizers.

5. Water logging has to be avoided.

6. Social forestry method can be adopted.

7. Organic farming practices can be beneficial.

Consumerism and Waste Products

Consumerism is the rights of people to acquire and use products and services. The sustainable consumerism leads to equitable distribution of resources.

In practice certain countries in the world consume major part and leaving a fraction to the rest of the world.

Globally, the 20% of the world's people in the highest-income countries account for 86% of total private consumption expenditures. The poorest 20% consume only a minuscule of 1.3%.

Sl. No.	Commodity	Consumption by Richest 5 Nations	Consumption by Poorest 5 Nations
1	Meet and fish	45%	5%
2	Energy	58%	4%
3	Telephone	74%	1.5%
4	Paper	84%	1.1%
5	Vehicle	87%	>1%

"Over" population is usually blamed as the major cause of environmental degradation, but the above statistics strongly suggests otherwise. As we will see, consumption patterns today are not to meet everyone's needs. The system that drives these consumption patterns also contribute to inequality of consumption patterns too.

Furthermore, the processes that lead to such disparities in unequal consumption are themselves wasteful and is structured deep into the system itself. Economic efficiency is for making profits, not necessarily for social good (which is treated as a side effect). The waste in the economic system is, as a result, deep.

Eliminating the causes of this type of waste are related to the elimination of poverty and bringing rights to all. Eliminating the waste also allows for further equitable consumption for all, as well as a decent standard of consumption.

So these issues go beyond just consumption and this section only begins to highlight the enormous waste in our economy which is not measured as such.

Waste Products

Those who consume more also produce more waste. Their contribution to environmental degradation is also high. Indeed countries which consume more must proportionately to contribute to environmental sustainability.

Advanced countries also generate more waste and degrade the environment due to technological advancements. Especially tons of electronic wastes are generated and dumped in the environment.

Consumerism is also associated with manufacturing technologies, raw material processing and bi-products generation. All these activities generate waste products which is deleterious to environment.

The emissions and effluents are also are proportionately high in countries with high consumer index. Buying power in such countries make them to discard the previous generation products and go for new versions.

Thus consumerism and waste products go side by side. So high consumer societies and countries must implement sustainable development concepts more rigorously to save the environment.

Further those who consume more should be more environment conscious and contribute more in environment preservation metnods.

Salient Features of Environment Protection Act, 1986

Objective: to the protection and improvement of environment and the prevention of hazards to human beings, other living creatures, plants and property.

1. The act has four chapters and 26 sections.
2. Chapter I defines the environment, environmental pollution, hazardous substance, etc.
3. Chapter II deals with steps to be taken to avoid environmental pollution. It also proposes to appoint appropriate authority with powers and function. It authorizes to make rules and regulation to prevent environment protection.
4. Chapter III describes the emission and effluent standards. It gives power enter and inspect and also take samples for analysis. It proposes the formation of environmental laboratories. It proposes five years imprisonment or upto 1 lakh rupees fine or both for any proved offence under the act. In subsequent offence 7 years of imprisonment and an additional fine of Rs.5,000/per day.
5. In case of government companies Head of the Department is liable for punishment.

6. Chapter IV protects person if action taken is in good faith without the intention to affect the environment. The offence under the act is not cognizable one. Court can act only if complaint is received.

Air (Prevention and Protection of Pollution) Act, 1981 (Act No. 14 of 1981)

Objective

An Act to provide for the prevention, control and abatement of air pollution for the establishment, of Boards, for conferring on and assigning to such Boards powers and functions.

To implement the decisions aforesaid in so far as they relate to the preservation of the quality of air and control of air pollution

Salient Features of the Act

1. The act has seven chapters and 54 sections.
2. Chapter I defines describes air pollution, air pollutants, emissions, etc.
3. Chapter II authorizes the constitution of Central and State Pollution Control Boards. It also directs separate Boards for Union Territories.
4. Chapter III prescribes the powers and functions of Central and State Pollution Control Boards. It gives power to enter industry, take samples, audit books, perform analysis, publish data, etc.
5. Chapter IV gives direction regarding prevention and control of air pollution. It gives power to give directions, restriction to use certain process, prescribing emission standards, etc.
6. Chapter V is with respect to found allocation, spending, accounts, audit and reporting
7. Chapter V prescribes penalty for offence under the act. Each proved failure not be less than one year and six months but which may extend to six years and with fine. In case the failure continues, with an additional fine up to five thousand rupees. The offence under the act is not a cognizable one.
8. Chapter VII describes the Powers of Central Government to make rules and also to dissolve State Pollution Control Board.

State Pollution Control Board under Air Act

Section 4 of Air (Prevention and Protection of Pollution) Act, 1981 authorises the constitution of State Pollution Control Board.

Constitution of State Pollution Control Board

There should be State Pollution Control Board.

- It is headed by a Chairman having a special knowledge and practical experience in the area.
- It should have members not exceeding 5 as the State Government may think fit.
- There should be non-official members, not exceeding 3, as the State Government may think fit to be nominated.
- Two persons to represent the companies controlled by State Government.
- A full-time member-secretary having such qualifications knowledge and experience of scientific, engineering or management

Functions of State Board

- Plan comprehensive programs and advice State Government to control air pollution.
- Coordinate with Central Board.
- Inspect pollution control equipments, industries, take necessary actions to control air pollution.
- Maintain control of emissions and impose standards.
- Give directions to Industries within the State.
- It is a body corporate can sue and can be sued.

Water (Prevention and Protection of Pollution) Act, 1974 (Act No. 6 of 1984)

Objectives

- For maintaining and restoring of wholesomeness of the water.
- For the establishment of the Boards with the powers and functions relating thereof and for the matters connected therewith.

Salient Features of Water Act

The act has four chapters and 18 sections.

- Chapter I defines outlet, pollution, sewage, effluent, trade effluent, etc.
- Chapter II describes the constitution of Central and State Pollution Control Boards.
- Chapter III describes the constitution of Joint Pollution Control Board.
- Chapter IV deals with the functions of the State Pollution Control Bards and Central Pollution Control Board

Constitution and Function of Central Pollution Control Board

The Central Board shall consist of the following members, namely

a) A full-time chairman, being a person having special knowledge or practical experience in respect to matters relating to environmental protection. He is nominated by the Central Government.

b) Members not exceeding five to be nominated by the Central Government.

c) Persons, not exceeding five to be nominated by the Central Government, from amongst the members of the State Boards.

d) Non-officials, not exceeding three to be nominated by the Central Government, to represent the interests of agriculture, fishery or industry or trade or any other interest.

e) Two persons to represent the companies or corporations owned, controlled or managed by the Central Government.

f) A full-time member-secretary, possessing qualifications, knowledge and experience of scientific, engineering or management.

g) The Central Board shall be a body corporate with the name aforesaid having perpetual succession and a common seal with power. It can sue can be sued.

Functions

a) Advise the Central Government on any matter concerning the prevention and control of water pollution

b) Co-ordinate the activities of the State Boards and resolve disputes among them;

c) Provide technical assistance and guidance to the State Boards

d) Plan and organise the training of persons engaged water and air pollution studies.

e) Organise through mass media a comprehensive program regarding the prevention and control of water and air pollution;

f) Collect, compile and publish technical and statistical data relating to water and air pollution.

Joint Pollution Control Board

Section 13 of Water Act, 1974 authorises to cosntittute Joint Pollution Control Board. .It shall consist of the following members, namely:--

A full-time chairman, being a person having special knowledge or practical experience in respect of matters relating to environmental protection to be nominated by the Central Government;

Two officials from each of the participating States to be nominated by the concerned participating State Government.

One person to be nominated by each of the participating State Governments from amongst the members of the local authorities functioning within the State concerned.

One non-official to be nominated by each of the participating State Governments to represent the interests or agriculture, fishery or industry or trade in the State concerned.

Two persons to be nominated by the Central Government to represent the companies or corporations owned, controlled or managed by the participating State Government.

A full-time member-secretary, possessing qualifications, knowledge and experience of scientific, engineering or management.

It performs functions referred ether by State Governments or Central Government.

Wildlife Protection Act, 1972 (Act No. 53 of 1972)

Objective

An Act to provide for the protection of Wild animals, birds and plants and for matters connected therewith or ancillary or incidental thereto.

Salient Features

- The act contains 66 Sections arranged in Seven Chapters and 6 Schedules.
- The Chapter I defines animal, animal articles, captive animals, closed area, dealer, habitat, hunting, sanctuary, etc.
- Chapter II describes appointment of authorities like Director, Assistant Director, Chief Wildlife Warden, etc, their powers and delegation. It also describes the constitution and duties of Wildlife Advisory Board.
- Chapter III prohibits the hunting of wild animals. and IIIA deals with hunting of wild animals and protection for plants. Chief Wildlife Warden can grant permission to hunt when the animal is beyond recovery, dangerous to public, dangerous to oneself, wounding or killing in good faith. However licence can be given to hunt for education, scientific research and scientific management.
- Chapter IIIA is with respect to protection of certain plant species. It regulates cutting, procession, cultivation, sales of specified plants. The Chief Wild Life Warden may with the previous permission of the State Government, grant to any person for the purpose of education, scientific research, collection, preservation and display in a herbarium of

any scientific institutions; or propagation by a person or an institution approved by the Central Government.

- Chapter IV deals with declaration of Santuaries, National Parks and Closed area. The Chief Wildlife Warden may, on application, grant to any person for purposes, investigation or study of wildlife and purposes ancillary or photography, scientific research, tourism, transaction of lawful business with any person residing in the sanctuary.
- Chapter IVA recognizes zoos and constitution and function of zoo authority.
- Chapter V is with respect to Trade or Commerce in Wild Animals, Animal Articles and Trophies. It declares wild animals are Government property. It prohibits dealing with trophies and animal articles without licence.
- Chapter VA Prohibition of Trade or Commerce in Trophies, Animal Articles, etc. derived from Certain Animals listed in the Schedule.
- Chapter VI deals with Prevention and Detection of Offences against animals and plants. Chief Wildlife Warden or the authorized officer or any forest officer or any police officer has the Power of entry, search, arrest and detention to issue a search warrant, to enforce the attendance of witiness, to compel the discovery and production of documents and material objects, and to receive and record evidence.
- Chapter VII deals with Miscellaneous Provisions. It protects actions taken in good faith. It gives power to State and Central Government to make or alter rules.

Penalties

The guilty person whoever contravenes the provisions of this Act shall be punishable with imprisonment for a term which may extent to three years or with fine which may extent Rs. 25,000 or with both.

Rights of Scheduled Tribes

Section 65 of this act says that nothing in this Act shall effect the hunting rights conferred on Scheduled Tribes of the Nicobar Islands in the Union Territory of Andaman & Nicobar Islands.

Wildlife Advisory Board

Section 6 of Wildlife Protection Act, 1972 authorizes the constitution of Wildlife Advisory Board.

Constitution

a. Minister in charge of Forest or Union Territory, or, Chief Secretary to the State Government, Union Territory hall be the Chairman

b. To members of the State Legislature

c. Secretary to the State Government, or the Government of the Union Territory, in charges of Forests;

d. The Forest Officer in charge of the State Forest Department, by whatever may be the designation.

e. An officer to be nominated by the Director;

f. Chief Wildlife Warden, ex-officio;

g. Officers of the State Forest Government not exceeding five

h. such other person, not exceeding ten, who, in the opinion of the State Government, are interested in the protection of Wildlife, including the representatives of tribals not exceeding three.

Duties of the Wildlife Advisory Board

It shall be the duty of the Wildlife Advisory Board to advise the State Government,–

a. In the selection of areas to be declared as Sanctuaries, National Parks, and Closed Areas and the administration there of

b. in formulation of the policy of protection and conservation of Wildlife and specified plants

c. in any matter relating to any schedule;

d. (cc) in relation to the measures to be taken for harmonizing the needs of the tribals and other dwellers of the forest with the protection and conservation of wildlife

e. in any matter that may be referred to it by the State Government.

Forest (Conservation) Act, 1980 (Central Act No. 69 of 1980) Amended in 1988

Objective

An Act to provide for the conservation of forests and for matters connected therewith or ancillary or incidental thereto.

Salient Features of the Act

- Prior approval of State or Central Government is required to use reserved forest for non-forest purposes, lease or otherwise to any private person, etc.

- Section 3 authorizes the constitution of Forest Advisory Committee. Its function is other matter connected with the conservation of forests which may be referred to it by the Central Government.
- When ever contravenes or abets the contravention of any of the provisions of the Act shall be punishable with simple imprisonment for a period which may extend to fifteen days.
- Section 4 authorizes the Central Government to make any rule with respect to forest conservation.

Forest Advisory Committee

Section 2A of the Amended Act, 1988 constitute the Advisory Committee with the following members.

1. Inspector General of Forests - Chairman, Ministry of Environment and Forests
2. Additional Inspector General of Forests – Member, Ministry of Environment and Forests
3. Joint Commissioner (Soil Conservation) – Member, Ministry of Agriculture
4. Three eminent environmentalists – Member (non-officials)
5. Deputy Inspector General of Forests - Member-Secretary (Forest Conservation), Ministry of Environment and Forests.

Functions

1. Committee to advise on proposals received by the Central Government if the area of the forest land involved is more than twenty hectares involving clearing of naturally grown trees in forest land.
2. Whether the forests land proposed to be used for non-forest purpose forms part of a nature reserve, national park wildlife sanctuary, biosphere reserve or forms part of the habitat of any endangered or threatened species of flora and fauna.
3. Whether the use of any forest land is for agricultural purposes or for the rehabilitation or persons displaced from their residences by reason of any river valley or hydro-electric project
4. Whether the State Government or the other authority has certified that it has considered all other alternatives and that no other alternatives in the circumstances are feasible and that the required area is the minimum needed for the purpose.
5. Whether the State Government or the other authority undertakes to provide at its cost for the acquisition of land of an equivalent area and afforestation thereof.

Biopiracy

Biopiracy refers to the privatization and unauthorized use of biological resources by entities (including corporations, universities and governments) outside of a country which has pre-existing knowledge. This privatization and use is sometimes claimed to be predatory. Particular activities usually covered by the term are:

- Exclusive commercial rights to plants, animals, organs, microorganisms, and genes;
- Commercialization of traditional communities' knowledge on biological resources;
- They are not commodities to be exploited by the corporate sector. Patenting life forms or knowledge is not acceptable. It is theft from generations past, present and future.

The story of Basmati

Basmati rice is distinct for its unique aroma and flavour. Basmati translates "queen of aroma". There is reference to Basmati in ancient texts, folklore and poetry, as it has been grown for centuries. There are twenty-seven distinct documented varieties of Basmati grown in India.

In 1997 RiceTec, a transnational corporation got patent rights on Basmati rice and grains, through the US Patent and Trademark Office. Rice Tec could have sole right to use the term 'Basmati' for marketing the rice anywhere in the world. If this patent is enforced internationally under World Trade Organisation rules it could seriously affect the livelihoods of many Indian and Pakistani farmers who grow and trade rice. Farmers could find themselves paying royalties on varieties their communities have played a hand in breeding over centuries.

Issues Involved in Enforcement of Environmental Legislation-Public Awareness

Poor Education: India is still blessed with uneducated and they know very little about what is environment even in local language.

Public Poverty: India is keeping its level of majority of the public in below poverty line due its excellent independent management over 60 years. As the poverty and environmentalism are opposite poles nothing will improve in terms of environment unless and until poverty alleviation and basic needs are satisfied.

Indian Culture: Indian tradition is grown with the such practices which are against environment. Indeed India forcibly forced to preserve Indian heritage in terms legal concessions. Right to live in the right environment of the ancestors is a fundamental right.

Gap Between Preaching and Practice: Those who preach at least practice or at least try to practice. Indeed the entire sandalwood and other trees cut down by Late Veerappan is utilized by elite groups. It appears the wild animals he killed fetched considerable foreign currency through some route. It was clearly argued that greatest polluter and anti-environmentalist is Indian Government. Right from Bhopal Tragedy to many Dams built under Government of India Supervision are questioned by intra and inter national level.

Technological Hesitance or Inevitability: India is in technological level under developing category. She is not in a position to keep up to the required level of competence for environmental safe guard. She provides shelter to many companies still to operate with out dated polluting technology with slight modification of having a pollution control and treatment plants which hardly does its duty.

Global Competition: India need to compete for economic sustainability. Though we make legal frame work in accordance with International requirements internally we have to sacrifice to a large extent of practicing the legal requirements. We do freely and fairly allow the pesticides like DDT to be manufactured and such contents like (very) soft drinks and also dye (die) industries and leather industries to perform their practices.

Laws with Loopholes: The laws have exceptions and special status and they generally and frequently invoked.

We are greater importer of environmental products for the purpose of foreign currency, thus law bends due to its maximum flexibility. Many times most of the environmental laws are over-ridded by provisions of other laws. For example lauding provisions of Forest

Conservation Act, Wild Life Protection Act can not stand near the Indian Railway Act. Thus, whenever wanted often and frequently wanted railway tracks forced or freely to go through dense forest preferably. India is a greatest importer of leather, leather products, bamboo products and all under respective protective acts.

Indian Public Authorities: We are blessed with right from beginning of Independence in all levels, flexible and friendly Public Servants and authorities, who can appropriately utilize the environmental laws and environmentalism.

Legislation and Public Awareness

We the Socialistic Democratic Constitution, people of India can contribute to a large extent for the preservation of environment.

- Every Indian citizen to be aware and made aware of the importance of environment.
- The legal and other social obligations of authorities.
- The citizen must adopt try to live in the direction of environmentally acceptable practices and avoid traditional ones.
- Public must see that any technological investment in terms of development should harm little to the environment.
- Reject which are harmful and accept the eco-friendly ones.
- Any utilization invoking provision of relevant law for agitation, protest, strike etc, should be no damage to the environment, though the authorities are adamant to hear the voice of concern.
- Discourage terrorism and report such activities to authorities to avoid damage to the ecosystem.
- Learn to live simple and eco-friendly manner and propagate the same.
- Abide the law of the State in letter and spirit, not to invoke exceptions.

Questions

1. What are causes for the present unsustainable equilibrium? Explain the methods adopted to make it sustainable?
2. What is a rain water harvesting and explain a simple method for it.
3. Explain the significance of watershed and how will you effectively manage it?
4. What is resettlement and rehabilitation? Explain government and non government sands in this area.
5. Explain global warming and its effects to environment.

6. Ozone layer depletion is due to air pollution-explain.

7. Nuclear accidents and nuclear holocaust wipes out human population and make

8. the environment non-sustainable-explain with case studies.

9. What is a wasteland, its reclamation and sound practices of waste land reclamation?

10. Explain consumerism and their right with respect to waste products.

11. What are the salient features of Environmental Protection Act?

12. Outline the different methods prescribed in air act to prevent air pollution.

13. How will you control water pollution using water act?

14. Wildlife protection act is meant for wildlife. How does it enforce laws under wildlife protection act?

15. Explain the essential features of Forest Conservation Act which conserves the forest ecology?

16. What are the problems in enforcing environmental law and how will you overcome it

Human Population, Health and Environment

Population growth, variation among nations – population explosion – family welfare programme – environment and human health – human rights – value education – HIV/AIDS – women and child welfare – role of information technology in environment and human health.

Population Growth

In 2000, the world had 6.1 billion human inhabitants. This number could rise to more than 9 billion in the next 50 years. For the last 50 years, world population multiplied more rapidly than ever before, and more rapidly than it will ever grow in the future.

On average per second 4-5 children born and two others die. The net gain is 2.5. Thus it is an addition of 9000 per hour, 2,14,000 per day and 78 million per year.

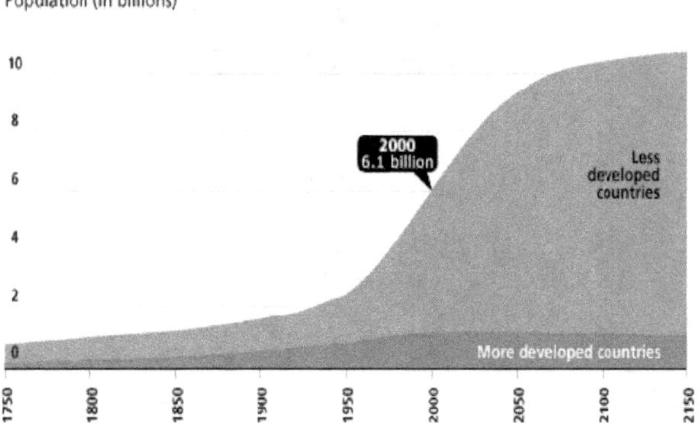

Theories of Population Growth

Malthusian Theory (1978)

Human population tends to increase at an exponential or compounded rate. The food production increase slowly or remain stable. The result is collapse into starvation, crime and misery.

Marxian Theory

Karl Marx view is population growth is symptom rather than the cause of poverty, resource depletion and social ills. The real cause is exploitation and oppression.

Characteristics of Population Growth

Three important type of population growth is notice among nations.

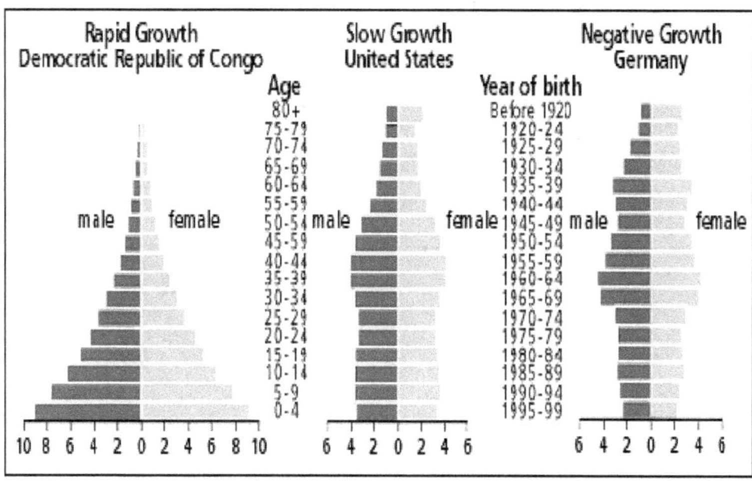

Exponential Growth: The growth rate is in the order 2,4,8,16,32, etc. The dramatic increase in population in the past 50 years is it following growth pattern.

$$N_t = N_0 e^{rt}$$

N_t = Number of human being after time t.

N_0 = Initial population size (when t = 0)

r = rate of growth

Slow Growth: There is a definite increase in population but not very drastic increase.

Negative Growth: The population size decreasing with respect to time. Negative growth is not due to scarcity of food or disease, etc but due to poor birth rate.

Determination of Growth Rate

$$\text{Growth Rate} = \frac{(\text{Population at end of a period} - \text{Population at the beginning})}{\text{Population at the beginning}}$$

Growth Ratio = Growth Rate x 100

The growth rate (ratio) is positive population is increasing and vice versa.

Zero Population Growth: If the number of birth plus immigration is equal to the number of death plus emigration then it is zero population growth.

Projections of World Population

No one really knows how large the world's population will be in the future. But we can make educated guesses by looking at past and present trends in two of the components of population growth: births and deaths. The third component, migration, can affect the growth of individual countries, but not to world population. World population is projected to increase to 7.8 billion by 2025, and to reach 8.9 billion by 2050.

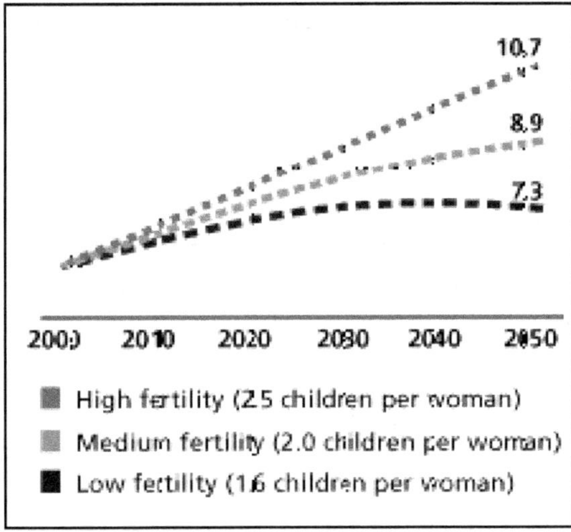

Total fertility rate (TFR): The average number of children delivered by a women during her childbearing years.

Doubling Time (Td): It is the time required to double the number of population.

$$Td = \frac{70}{R} \qquad R = \text{annual Growth Rate}$$

For India annual growth rate is about 2 (1.92). So doubling time is about 35 years.

Replacement Level: It is the replacement of parents by children. If two children in a family the parents (father and mother) are replaced by them. If infant mortality is high then it is not a simple calculation.

Infant Mortality Rate: It is the ratio of number of children born and died in one year.

Population Stabilization Ratio: It is the ratio of birth rate to death rate. It is near one for developed countries and more than one for others.

Demographic Transition: It is a model used to explain the correlation of population variation with economic development of a country. As the economic growth is high then decrease in birth rate and death rate is observed. This phenomenon is referred as demographic transition.

Carrying Capacity: The maximum population that can be supported by an ecosystem is known as carrying capacity. When it exceeds this value ecological imbalance starts and scarcity is the consequence.

Variation among Nations

The world population is the total number of human being living on earth on time. On December, 2008 is is just over 6.725 billion. The present population growth rate is about 1.9 percent.

In 1800, the vast majority of the world's population (86 percent) resided in Asia and Europe, with 65 percent in Asia alone (see chart, "World population distribution by region, 1800-2050"). By 1900, Europe's share of world population had risen to 25 percent, fueled by the population increase that accompanied the Industrial Revolution. Some of this growth spilled over to the Americas, increasing their share of the world total. The closer trend continued in 2000 also.

Between 2000 and 2030, nearly 100 percent of this annual growth will occur in the less developed countries in Africa, Asia, and Latin America, whose population growth rates are much higher than those in more developed countries. Growth rates of 1.9 percent and higher mean that populations would double in about 36 years, if these rates continue.

The populations in the less developed regions will most likely continue to command a larger proportion of the world total. While Asia's share of world population may continue to hover around 55 percent through the next century, Europe's portion has declined sharply and could drop even more during the 21st century. Africa and Latin America each would gain part of Europe's portion. By 2100, Africa is expected to capture the greatest share (see chart, "World population distribution by region, 1800-2050").

The more developed countries in Europe and North America, as well as Japan, Australia, and New Zealand, are growing by less than 1 percent annually. Population growth rates are negative in many European countries, including Russia (-0.6%), Estonia (-0.5%), Hungary (-0.4%), and Ukraine (-0.4%).

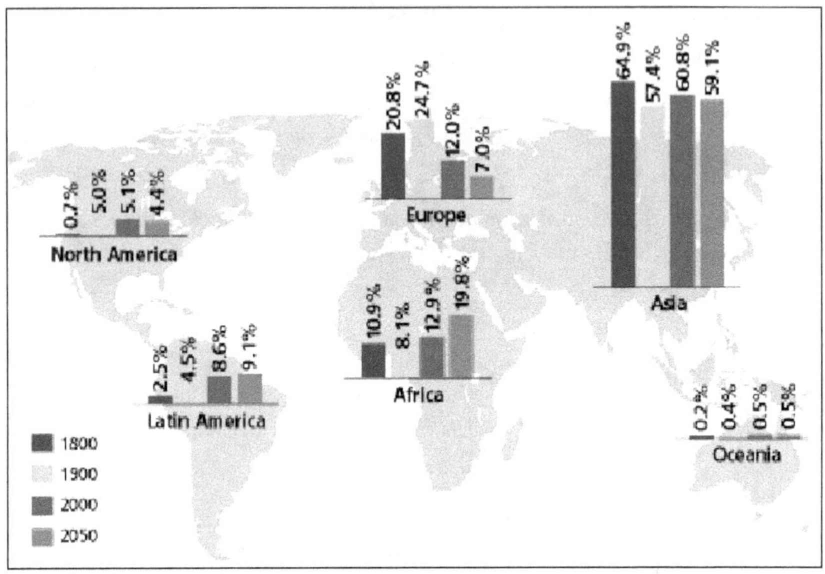

Reason for Variation

Population change results from the interaction of three variables: births, deaths, and migration. This relationship is summarized by a formula known as the balancing equation. The difference between births and deaths in a population produces the natural increase (or decrease) of a population. Net migration is the difference between the number of persons entering a geographic area (immigrants) and those leaving (emigrants).

$$\begin{pmatrix} \text{Births - Deaths} \\ \text{or Natural Increase} \end{pmatrix} + \begin{pmatrix} \text{Immigrants - Emigrants} \\ \text{or Not Migration} \end{pmatrix} = \begin{pmatrix} \text{Growth (or)} \\ \text{Decrease} \end{pmatrix}$$

In addition climatic condition greatly varies. It has the influence on human fertility rate. In addition developed countries are concentrating on technological development. The developing and less developed countries are more on family building and care.

In the present scenario from the graph appears to be in developed countries birth and death rate is closer and thus stabilized. In less developed countries birth rate is much higher than death rate.

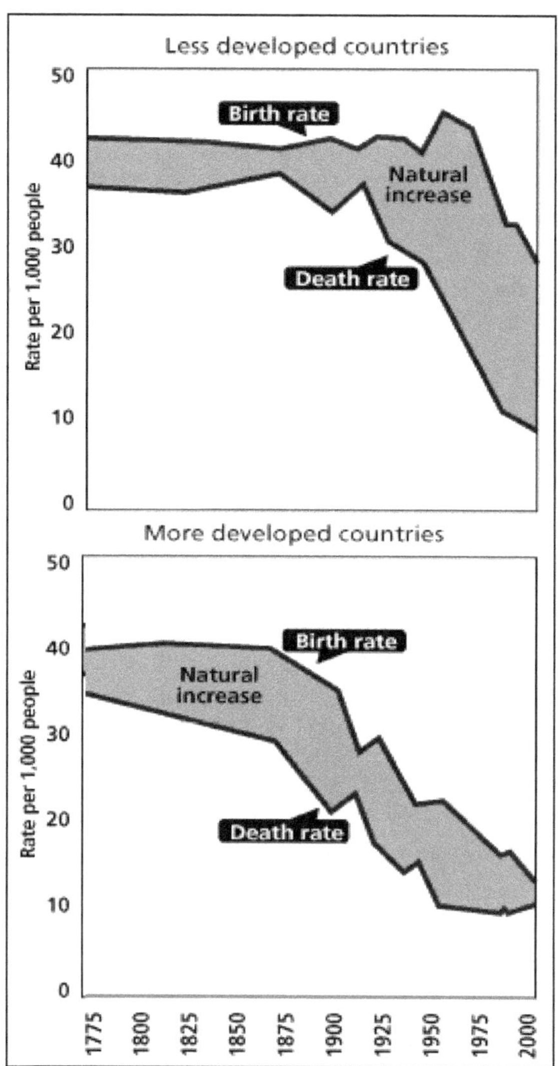

Population Explosion

It refers to rapid (exponential) increase in human population beyond the carrying capacity of the ecosystem. It is not a function of size and density of the population in a given area rather correlation with environment and sustainability.

Indeed counties with very low population but with rich resources exponential growth are good and positive. However if it is in countries which are already scare resources becomes dangerous.

World population growth accelerated after World War II, when the population of less developed countries began to increase dramatically. After millions of years of extremely slow growth, the human population indeed grew explosively, doubling again and again; a billion people were added between 1960 and 1975; another billion were added between 1975 and 1987. Throughout the 20th century each additional billion has been achieved in a shorter period of time. Human population entered the 20th century with 1.6 billion people and left the century with 6.1 billion.

The growth of the last 200 years appears explosive on the historical timeline. The overall effects of this growth on living standards, resource use, and the environment will continue to change the world landscape long after.

It was not until the 1700s that the modern era of population growth began. Between 1850 and 1900, the annual growth rate reached 0.5 percent. The rate surged to 2.2 percent by the mid-1960s, dropped to 2 percent by the mid-1980s, and declined to about 1.9 percent by 2000.

Reason for Rapid Population Growth

- More matured after practically demonstrating by different wars. Thus avoided war, peace become more prevalent.
- Administration reforms like political system replaced small Kingdoms and thus fight for territory and superiority leading to life loss is reduced.
- More heath care options introduced. Many of the contagious disease are brought under control or completely eliminated. Thus decrease in death rate contributed to population growth.
- Need for more children are realized since it is considered as wealth to the family.
- The average life time of human being also increased substantially. It was calculated to be 40 years during 1950's. The present life expectancy is about 61 years.
- The development in medical field and chemotherapy decreased the child mortality rate and also labour becomes easier. Thus women feel happy to carry many children.
- Religious and social faith also contributed for population explosion in counties like India.

Consequences of Population Explosion

- **Environmental degradation due to overexploitation even for sustainable life.**
- Unhealthy competition leading to problems in law and order problem.
- Desertion and careless population causing administrative problems.
- Food scarcity and poverty.
- Increase in crime rate.
- Spread of diseases like HIV/AIDS due more number of sex worker and flesh trade.

The Family Welfare Program

A high degree care given to families by Government of India, Non-Government Organizations and Service Organizations and Individuals comes under family welfare programs.

Ministry of Health & Family Welfare Department of Family Welfare are the Government machinery operating in India.

Objectives

- Family is the basic unit of human population and it health is certainly country's health.
- Due to the presence of large majority of family in India in poverty they can not afford to have health care needs.
- Illiteracy and poor communication system warrant the need to educate the society in social obligations.
- Environmental sustainability and population stabilization rests in the hands of family those who are in rural and illiterates.

Need for Family Welfare Program

- International obligations and Human Right requirements
- Constitutional Obligation
- Humanitarian perspective
- Social Obligation to those who needs and their bare minimum necessity.

Basic Health Care Services

A list of services which comprehensives primary (or basic) care should includes the following

- General practitioner / family physician services for personal health care.

- First level referral hospital care and basic specialist services - pediatrics, gynecology and obstetrics, general medicine, general surgery, dental services and ophthalmology, including special diagnostics, like polio, tuberculosis, etc.
- Immunization services for vaccine preventable diseases
- Maternity services for safe pregnancy, abortion, delivery and postnatal care
- Pharmaceutical services - supply of only rational and essential drugs as per accepted standards.
- Epidemiological services, including laboratory services, surveillance and control of major diseases with the aid of continuous surveys, information management and public health measures.
- Contraceptive services.
- Health education and information.
- Ambulance services.

Child Welfare Programs

Integrated Child Development Scheme: The Scheme was started in the year 1975 as a pilot project with the following objectives

- To improve the health and nutritional status of children in the age group of 0-6 years.
- To let the foundation of proper psychological, physical and social development of the children.
- To reduce the incidents of mortality, morbidity, and malnutrition and school drop outs.
- To achieve effective co-ordination of policy and integration amongst various departments to promote child development.

- To enhance the capability of mother to look after the normal health and nutritional need of the children through proper nutrition and health education.

To cover the above objectives the following package of services were also introduced :

1. Early Childhood Education (ECE)
2. Supplementary Nutrition Programme
3. Immunization
4. Health Check-up
5. Nutrition and Health Education
6. Referral Services

Child Labour in India

Poor children in India begin working at a very young and tender age. Many children have to work to help their families. Some families expect their children to continue the family business at a young age. India has always stood for constitutional, statutory and developmental measures that are required to eliminate child labour in India. The Child Labour (Prohibition and Regulation) Act 1986 regulates the child labourers. Right Education (Article 21A) has been introduced in Constitution but yet to come into force.

Cradle Baby Scheme

The Cradle Baby Scheme was launched in Salem in the year 1992 with the aim of eradicating female infanticide. This Scheme was later extended during 2001 to Madurai, Theni, Dindigul and Dharmapuri, Till 31.3.2008, this program has saved the life of 3,044 female children.

Women Welfare Programs

National Commission for Women, a Government of India body to specifically look after women welfare. Women and Children are protected by Constitution (Article 23, 24). Dowry Prohibition Act, Domestic Violence Act, Maternity Relief Act, etc further strengthen hands of Women in society. International Convention on the Elimination of All Forms of Discrimination Against Women protects women.

Family Planning Programs

By considering unsustainable growth of Indian population Department of Family Welfare takes serous efforts to bring the growth of population under control. It provides family education and services. It gives also services in birth control.

Permanent Birth Control Methods

(a) **Tubectomy:** It is permanent sterilization of females from being pregnant. Laparoscopes/Laparocators are sterilization which is a relatively quicker method of female sterilization.

(b) **Vasectomy:** It is the permanent sterilization of males. No-Scalpel Vasectomy (NSV) is one of the most effective contraceptive methods available for males.

Temporary Birth Control Methods

(i) **Condoms:** It is male birth control device. In 2003, there are about 14 million condoms are distributed. Condoms also avoids the spread of sexually transmittable diseases.

(ii) **Copper-T (Cu-T):** It is a female birth control device. It is used for spacing between pregnancies.

(iii) **Oral Contraceptives:** It is to prevent pregnancy used by female partners. Mala D, Sugam, Suvida, Khushi, etc are prevalent and in the market also supplied by health care department. It 2003 about 10 lakh oral contraceptives are distributed. Saheli is a weekli oral contraceptive for women. Levonorgentrel is a Emergency Contraceptive is to be taken on prescription of Medical Practitioners to prevent unwanted pregnancy after unprotected act of sexual intercourse.

Vaccination Programs

Under the Immunization Programme, Department supplies six vaccines for vaccinations to infants and pregnant women for the control of vaccine preventable diseases.

- Tatanus Toxoid Vaccine
- DT Vaccine
- OralPolioVaccine
- Measles Vaccine
- BCG Vaccine
- DPT Vaccine

The Pre-Natal Diagnostic Techniques (PNDT)

The pre-natal diagnostic techniques like amniocentesis and sonography are useful for the detection of genetic or chromosomal disorders or congenital malformations or sex linked disorders, etc.

Though it is misused for sex determination, the centers can give effective advice on the health of child in the womb. Sex determination is crime and punishable under Pre-Natal Diagnostic Techniques (Regulation and Prevention of Misuse) Act of 1994 subsequently amended into the PCPNDT Act in 2003.

Termination of Pregnancy

The Medical Termination of Pregnancy Act, 1971 Section 3(1) reads as "Notwithstanding anything contained in the Indian Penal Code a registered medical practitioner shall not be guilty of any offence under that Code or under any other law for the time being in force, if any pregnancy is terminated by him in accordance with the provisions of this Act".

Working Women's Hostel

The Working Women, particularly from the middle and lower middle classes, are facing many problems to find suitable accommodation at the fees affordable to them. Government of Tamil Nadu has allotted land and found for this purpose in the recent budget.

Environment and Human Health

The environment where human lives and the health status he receives are interlinked. Clean environment is required for the safe and comfortable life. The importance of environment with respect to human health can be understood from the following table.

Sl. No.	Environment	Consequence
1	Pesticides	Parkinson's disease
2	Air Pollution	Asthma, allergy and skin problems, cancer
3	Tobacco smoke	Lung cancer
4	Water pollution with microbes	Cholera, Jaundice, etc
5	Water pollution with chemicals	Cancer, reproductive problems, nervous disorders
6	Food Pollution	Diabetes, obesity, cholera.
7	Radiation	Reproductive disorders, eye defect, skin problems
8	Metals poisoning	Skin problems, birth defects, nervous disorders
9	Climate Change	Growth problems, reproductive disorders.

Thus every effort has to be made to keep the environment clean and healthy. If any deviations or interruptions it not only affects the ecological balance but also human health.

Not In My Back Yard (NIMBY) Syndrome

It is the common perception of the population that the pollution is not in and around me. So why should I worry about it.

One should be clear pollution or problem any is the problem everywhere. Indeed ecosystem is an integrated system. Any imbalance in one ecosystem the consequence is carried over to entire globe.

Human Rights

Whereas disregard and contempt for human rights have resulted in barbarous acts which have outraged the conscience of mankind, and the advent of a world in which human beings shall enjoy freedom of speech and belief and freedom from fear and want has been proclaimed as the highest aspiration of the common people, Whereas it is essential, if man is not to be compelled to have recourse, as a last resort, to rebellion against tyranny and oppression, that human rights should be protected by the rule of law...

These are the second and third paragraphs of the preamble to the Universal Declaration of Human Rights (UDHR), adopted by the United Nations General Assembly on December 10, 1948 without a dissenting vote.

The United Nations Charter, Universal Declaration of Human Rights, and UN Human Rights convenants were written and implemented in the aftermath of the Holocaust, revelations coming from the Nuremberg war crimes trials.

Every gun that is made, every warship launched, every rocket fired signifies in the final sense, a theft from those who hunger and are not fed, those who are cold and are not clothed. This world in arms is not spending money alone. It is spending the sweat of its laborers, the genius of its scientists, the hopes of its children. This is not a way of life at all in any true sense. Under the clouds of war, it is humanity hanging on a cross of iron.

-- Dwight D. Eisenhower

A Short History of the Human Rights Movement

- The King John of England had given right of the church to be free from governmental interference, the rights of all free citizens to own and inherit property and be free from excessive taxes.

- In the eighteenth and nineteenth centuries in Europe several philosophers proposed the concept of "natural rights," rights belonging to a person by nature and because he was a human being.

- In 1789 the people of France overthrew their monarchy and established the first French Republic. Out of the revolution came the "Declaration of the Rights of Man."

- Thoreau is the first philosopher to use the term, "human rights", and does so in his treatise, *Civil Disobedience*. This work has been extremely influential on individuals as Mahatma Gandhi, and Martin Luther King.

- Other early proponents of human rights were English philosopher John Stuart Mill, in his *Essay on Liberty* and American political theorist Thomas Paine in his essay, *The Rights of Man*.

- The middle and late 19th century saw a number of issues slavery, serfdom, brutal working conditions, starvation wages, child labor, etc. The concept "all men are created equal" evolved.

- In 1961 a group of lawyers, journalists, writers, and others, offended and frustrated by the sentencing of two Portugese college students to twenty years in prison for having raised their glasses in a toast to "freedom" in a bar.

United Nations in Pursuit of Human Right

The UN recognizes a wide range of human rights issues.

These include a definition of indigenous peoples, the role of intergovernmental and non-governmental organizations, the elimination of discrimination, and basic human rights principles, as well as special areas of action in fields such as health, housing, education, language, culture, social and legal institutions, employment, land, political rights, religious rights and practices, and equality in the administration of justice. On December 10, 1948 the General Assembly of the United Nations adopted and proclaimed the Universal Declaration of Human Rights.

Preamble

Whereas recognition of the inherent dignity and of the equal and inalienable rights of all members of the human family is the foundation of freedom, justice and peace in the world,

Whereas disregard and contempt for human rights have resulted in barbarous acts which have outraged the conscience of mankind, and the advent of a world in which human beings shall enjoy freedom of speech and belief and freedom from fear and want has been proclaimed as the highest aspiration of the common people,

Whereas it is essential, if man is not to be compelled to have recourse, as a last resort, to rebellion against tyranny and oppression, that human rights should be protected by the rule of law,

Whereas it is essential to promote the development of friendly relations between nations,

Whereas the peoples of the United Nations have in the Charter reaffirmed their faith in fundamental human rights, in the dignity and worth of the human person and in the equal rights of men and women and have determined to promote social progress and better standards of life in larger freedom,

Whereas Member States have pledged themselves to achieve, in co-operation with the United Nations, the promotion of universal respect for and observance of human rights and fundamental freedoms.

Where as a common understanding of these rights and freedoms is of the greatest importance for the full realization of this pledge,

Now, Therefore THE GENERAL ASSEMBLY proclaims THIS UNIVERSAL DECLARATION OF HUMAN RIGHTS as a common standard of achievement for all peoples and all nations, to the end that every individual and every organ of society, keeping this Declaration constantly in mind, shall strive by teaching and education to promote respect for these rights and freedoms and by progressive measures, national and international, to secure their universal and effective recognition and observance, both among the peoples of Member States themselves and among the peoples of territories under their jurisdiction.

Articles	Constitution	Description
Article 1		All human beings are born free and equal in dignity and rights. They are endowed with reason and conscience and should act towards one another in a spirit of brotherhood.
Article 2	No Discrimination (Article 15)	Everyone is entitled to all the rights and freedoms set forth in this Declaration, without distinction of any kind, such as race, colour, sex, language, religion, political or other opinion, national or social origin, property, birth or other status. Furthermore, no distinction shall be made on the basis of the political, jurisdictional or international status of the country or territory to which a person belongs, whether it be independent, trust, non-self-governing or under any other limitation of sovereignty.
Article 3	Right to Life (Article 21)	Everyone has the right to life, liberty and security of person.
Article 4	Free from exploitation (Article 23)	No one shall be held in slavery or servitude; slavery and the slave trade shall be prohibited in all their forms.
Article 5		No one shall be subjected to torture or to cruel, inhuman or degrading treatment or punishment.
Article 6		Everyone has the right to recognition everywhere as a person before the law.
Article 7	Equality before law (Article 14)	All are equal before the law and are entitled without any discrimination to equal protection of the law. All are entitled to equal protection against any discrimination in violation of this Declaration and against any incitement to such discrimination.
Article 8	Right to legal remedy (Article 22)	Everyone has the right to an effective remedy by the competent national tribunals for acts violating the fundamental rights granted him by the constitution or by law.
Article 9		No one shall be subjected to arbitrary arrest, detention or exile.
Article 10		Everyone is entitled in full equality to a fair and public hearing by an independent and impartial tribunal, in the determination of his rights and obligations and of any criminal charge against him.
Article11 **(1)** **(2)**		Everyone charged with a penal offence has the right to be presumed innocent until proved guilty according to law in a public trial at which he has had all the guarantees necessary for his defense. No one shall be held guilty of any penal offence on account of any act or omission, which did not constitute a penal offence, under national or international law, at the time when it was committed. Nor shall a heavier penalty be imposed than the one that was applicable at the time the penal offence was committed.
Article 12		No one shall be subjected to arbitrary interference with his privacy, family, home or correspondence, nor to attacks upon his honour and reputation. Everyone has the right to the protection of the law against such interference or attacks.
Article 13 **(1)** **(2)**	Freedom of movement (Article 19 (1d))	Everyone has the right to freedom of movement and residence within the borders of each state. Everyone has the right to leave any country, including his own, and to return to his country.
Article 14 **(1)** **(2)**		Everyone has the right to seek and to enjoy in other countries asylum from persecution. This right may not be invoked in the case of prosecutions genuinely arising from non-political crimes or from acts contrary to the purposes and principles of the United Nations.
Article 15 **(1)** **(2)**	Right to citizenship (Article 19(1d))	Everyone has the right to a nationality. No one shall be arbitrarily deprived of his nationality nor denied the right to change his nationality.
Article 16 **(1)** **(2)** **(3)**		Men and women of full age, without any limitation due to race, nationality or religion, have the right to marry and to found a family. They are entitled to equal rights as to marriage, during marriage and at its dissolution. Marriage shall be entered into only with the free and full consent of the intending spouses. The family is the natural and fundamental group unit of society and is entitled to protection by society and the State.
Article 17 **(1)** **(2)**	Right to property (Repealed)	Everyone has the right to own property alone as well as in association with others. No one shall be arbitrarily deprived of his property.
Article 18	Freedom of religion (Artilce 26)	Everyone has the right to freedom of thought, conscience and religion; this right includes freedom to change his religion or belief, and freedom, either alone or in community with others and in public or private, to manifest his religion or belief in teaching, practice,

		worship and observance.
Article 19	Freedom of Expression (Article (19(1a))	Everyone has the right to freedom of opinion and expression; this right includes freedom to hold opinions without interference and to seek, receive and impart information and ideas through any media and regardless of frontiers.
Article 20 **(1)** **(2)**	Freedom of peaceful assembly (Article (19b)	Everyone has the right to freedom of peaceful assembly and association. No one may be compelled to belong to an association.
Article 21 **(1)** **(2)** **(3)**		Everyone has the right to take part in the government of his country, directly or through freely chosen representatives. Everyone has the right of equal access to public service in his country. The will of the people shall be the basis of the authority of government; this will shall be expressed in periodic and genuine elections which shall be by universal and equal suffrage and shall be held by secret vote or by equivalent free voting procedures.
Article 22		Everyone, as a member of society, has the right to social security and is entitled to realization, through national effort and international co-operation and in accordance with the organization and resources of each State, of the economic, social and cultural rights indispensable for his dignity and the free development of his personality.
Article 23 **(1)** **(2)** **(3)** **(4)**	Freedom of Profession (Article 19(1g))	Everyone has the right to work, to free choice of employment, to just and favourable conditions of work and to protection against unemployment. Everyone, without any discrimination, has the right to equal pay for equal work. Everyone who works has the right to just and favourable remuneration ensuring for himself and his family an existence worthy of human dignity, and supplemented, if necessary, by other means of social protection. Everyone has the right to form and to join trade unions for the protection of his interests.
Article 24		Everyone has the right to rest and leisure, including reasonable limitation of working hours and periodic holidays with pay.
Article 25 **(1)** **(2)**	Right to food and good health	Everyone has the right to a standard of living adequate for the health and well-being of himself and of his family, including food, clothing, housing and medical care and necessary social services, and the right to security in the event of unemployment, sickness, disability, widowhood, old age or other lack of livelihood in circumstances beyond his control. Motherhood and childhood are entitled to special care and assistance. All children, whether born in or out of wedlock, shall enjoy the same social protection.
Article 26 **(1)** **(2)** **(3)**	Right to education (Article 21A)	Everyone has the right to education. Education shall be free, at least in the elementary and fundamental stages. Elementary education shall be compulsory. Technical and professional education shall be made generally available and higher education shall be equally accessible to all on the basis of merit. Education shall be directed to the full development of the human personality and to the strengthening of respect for human rights and fundamental freedoms. It shall promote understanding, tolerance and friendship among all nations, racial or religious groups, and shall further the activities of the United Nations for the maintenance of peace. Parents have a prior right to choose the kind of education that shall be given to their children.
Article 27 **(1)** **(2)**	Right to culture (Article 26)	Everyone has the right freely to participate in the cultural life of the community, to enjoy the arts and to share in scientific advancement and its benefits. Everyone has the right to the protection of the moral and material interests resulting from any scientific, literary or artistic production of which he is the author.
Article 28		Everyone is entitled to a social and international order in which the rights and freedoms set forth in this Declaration can be fully realized.
Article 29 **(1)** **(2)** **(3)**		Everyone has duties to the community in which alone the free and full development of his personality is possible. In the exercise of his rights and freedoms, everyone shall be subject only to such limitations as are determined by law solely for the purpose of securing due recognition and respect for the rights and freedoms of others and of meeting the just requirements of morality, public order and the general welfare in a democratic society. These rights and freedoms may in no case be exercised contrary to the purposes and principles of the United Nations.
Article 30		Nothing in this Declaration may be interpreted as implying for any State, group or person any right to engage in any activity or to perform any act aimed at the destruction of any of the rights and freedoms set forth herein.

India and Human Right

Human Rights, "Human rights touch every right that a human being can possibly think of ... They are not gifts of the Universal Declaration of Human Rights or the Constitution but are only catalogued in them," NHRC Chairperson A S Anand said addressing a conference of state Chief Secretaries and Directors General of Police organized by the National Human Rights Commission here.

Human Rights are embedded in Part III of Indian Constitution as Fundamental Rights.

Value Education

Value: Value on human is a perception being as human not because but because actions and behaviours. Value is an asset to society. Values come by practice not by mere reading or seeing. Value gives belief and confidence to others.

Education: Education is not merely a matter of training the mind. Training makes for efficiency, but it does not bring about completeness. Education is not merely acquiring knowledge, gathering and correlating facts; it is to see the significance of life as a whole.

***EDUCATION is the Manifestation of the Perfection Already in Man...*Swami Vivekananda.**

Value and Education

Education should help us to discover lasting values. The purpose of education is not to produce mere scholars, technicians and job hunters, but integrated men and women who are free of fear; for only between such human beings can there be enduring peace.

Education without value is dangerous but a value added person without education is indeed valuable. Life, health and happiness rest on value not on education. Value brings happiness, education adds distress and insecurity. Value adds real success, education gives momentary gains. Simplicity is associated with value. Complications and complexity is product of education.

Value and Environment

It is clear from the hundreds of education, that education without value destroyed nature, human and other life forms. Seemingly limitless degradations done on environment on unsustainable manner worsened the health and happiness of life forms.

Instances such as nuclear war, degradation of forest, use of tons of explosives in the pretext of mining, vast industrial complexes in the name of business economy, etc are all due the outstanding education without value in it.

Education brought lifestyles, conveniences without comfort, enhanced population without brotherhood, humanity, etc. Human population with education without human values is indeed equivalent no human beings at all. Thus there must be value on human beings in particular on educated human beings.

Attach values on learning, respect the surrounding, environment, people, culture society and thus ends in value education. In such education a sustainable lifestyle embedded in it. Such individuals respect oneself and environment.

Great emperor Asoka is known for his value not because of his mightiness in war. The discoverer of atom bomb is least known than Sami Vivekanada only because of value.

A National Resource Centre for Value Education (NRCVE) has been recently set up in the Department of Educational Psychology and Foundations of Education (DEPEE) with the objective of developing and implementing program and strategies to renew emphasis on value inculcation and develop relevant material for use by various groups. Head, DEPEE is the coordinator of the Centre. Thus value education is realized at the highest level.

Value Education in Curriculum

1. First define the definite values in the curriculum.
2. Values have to be incorporated in the minds of human in childhood itself.
3. It has to be implemented in a definite tangible form such that to know that the value is worthy.
4. Practical approaches like community development programs, environment awareness programs so that relevance of value is implemented.

5. Life and achievements of value based persons has to be taught and moral has to be explored to them.

6. Children have to be learned think is social behaviour rather than economy, money, etc.

7. Not to make them quick achievers rather encourage them to understand the importance of life and living.

8. Make them to think before every activity whether environmentally sustainable or socially acceptable in a long term perspective.

HIV/AIDS

Human Immunodeficiency Virus (HIV) is responsible for the disease responsible for Aquired Immuno Difficiency Syndrome (AIDS). It is a sexually transmitting diseases. It breaks down human immune system by suppressing T-cells (lymphocytes). The 1st of December is marked as world aids day. HIV/AIDS reaches into every corner of society, affecting parents, children and youth, teachers and health workers, rich and poor.

Symbol of AIDS

The Red Ribbon is the global symbol for solidarity with HIV positive and people living with AIDS and it unites the people in the common fight against this disease.

The Red Ribbon is:

- Red like love, as a symbol of passion and tolerance towards those affected.

- Red like blood, representing the pain caused by the many people that died of AIDS.

- Red like the anger about the helplessness by which we are facing a disease for which there is still no chance for a cure.

- Red as a sign of warning not to carelessly ignore one of the biggest problems of our time.

Origin of HIV

Human terror activity on microbial population by using antibiotics, vaccines, etc is believed to be the main cause of HIV transformed virus. Since the survival become dangerous, they adopted a new mechanism of propagation by long dormancy followed by quick mutation with out loosing disease causing ability. It appears initial research and execution was done on African monkey, Chimpanzee. Then invaded human through vaccines developed from the monkey.

Reason for Concern

Due to high degree of tolerance and frequent mutation, the usual antibiotics and vaccine technology can not be a therapeutic method. Almost all chemotherapy fails to act on the virus. Intelligently destroys the human immune system and white blood cells as the first target. Thus natural tolerance and immune system is helpless. So other diseases also accompanied by HIV infection. Since it is propagating mostly through sexual intercourse, the most usual/casual enjoyment is in great trouble.

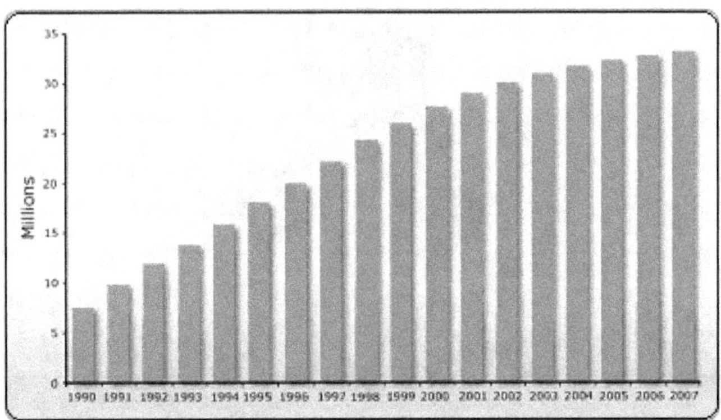

Gravity of the Problem

More than 25 million people have died of AIDS since 1981. Africa has 11.6 million AIDS orphans. Young people (under 25 years old) account for half of all new HIV infections worldwide.

It apears India is having about 5.7 million AIDS patients. But the recent report says it is only about 3 million if so accupies the second position in the world. Maharastra stands first and Tamil Nadu follows in total number of AIDS cases among States. India has a large population and population density, low literacy levels and consequently low levels of awareness. So HIV/AIDS is one of the most challenging public health problems ever faced by the country.

Sl.No.	Population Group	Population (million)
1	Adults living with HIV/AIDS in 2007	30.8 million
2	Women living with HIV/AIDS in 2007	15.5 million
3	Children living with HIV/AIDS in 2007	2.0 million
4	People newly infected with HIV in 2007	2.7 million
5	Children newly infected with HIV in 2007	0.37 million
6	AIDS deaths in 2007	2.0 million
7	Child AIDS deaths in 2007	0.27 million

Mechanism of Action of HIV

The HIV virus when it enters the human body politely occupies inside the white blood cells (WBC). It live upto 10 years silently and teaches WBC to manufacture HIV protein only and thus its immune response is destroyed. Large scale virus proteins come out from WBC and infect entire body and thus the condition called AIDS appears.

Diagnosis of HIV

Enzyme Linked Immuno Sorbant Assay (ELISA) is used for the diagnosis of HIV affected individuals. Blood serum is used for this purpose. Voluntary organizations and Government agencies performs this test and the information about the individual is kept confidential.

Symptoms

Prolonged fewer, diarrhea, caugh, tuberculosis, fall of hair from the head, loss of body weight, skin patches especially in legs and hands, etc.

Preventive Measures

Protected Sex: Follow the family culture of India. With strangers use condoms before having sexual intercourse.

Blood Transmission: Only after through checking blood transmission has to be accepted.

Organ Transplantation: Accept organ/tissue transplantation only with reliable and tested persons.

Injecting Needles: Not to use the once used needle. Always use disposable needle for medical injections.

Education: Educate the public and sex workers about the consequences and safety methods.

Medical Treatment: Take medical advice in worst condition. Azidothymidine (AZT) is the prescription drug, though it works only in early stages of AIDS. Now combination therapy is followed with limited success.

Vulnerable Groups

Migrants

Sex workers

Injecting Drug Users (IDUs)

Truck drivers

Honosexuals, Lesbians, etc.

The Early Years of the Response to HIV/AIDS in India

The first case of HIV infection in India was diagnosed among commercial sex workers in Chennai, Tamil Nadu in 1986. A number of screening centres were established throughout the country. Initially, the focus was on screening foreigners, especially foreign students. Then focused on screening blood banks.

A National AIDS Control Program was launched in 1987 with the program activities covering surveillance, screening blood and blood products and health education. In 1992 the National AIDS Control Organization (NACO) was established. NACO carries out India's National AIDS Program, which includes the formulation of policy, prevention and control programs.

The Government launched a Strategic Plan for HIV/AIDS prevention under the National AIDS Control Project. The Project established the administrative and technical basis for program management and also set up State AIDS bodies in 25 states and 7 union territories. The Project was able to make a number of important improvements in HIV prevention such as improving blood safety.

Legislation and Policies

Based on International and National law HIV affected persons can not be discriminated in Public Employment and isolated.

Role of Information Technology in Environment and Human Health

Uses of Information Technology

- Information technology has information as database regarding environment. It also receives satellite information and stores and dissimilate to public.

- Agriculture database can be used for crop and land management practices.

- It can predict natural calamities in coordination with sateliites.

- It can help in mitigation processes.

- Geological information, ocean currents, climate change data, etc can be obtained from IT database.

- National Management Information System (NMIS) has the database R & D Projects and Scientists working on it.

- Environment Information System (ENVIS) generates and maintain database on pollution, clean technology, remote sensing, biodiversity, desertification, environment management, etc.

- Geographical Information System (GIS), in coordination with satellite maintains the information regarding earth's water resource, soil type, cropland, polluted zones, land degradation, etc.

IT and Environment

- Continuously meddles the environment by ewastes.
- Due to criminal virus many storage units are discarded. The metallic impurities pollute the environment.
- Rapid advances in the field makes the hardware and accessories obsolete and make the earth as dumping ground.
- IT data transmissions in various area is advisable especially to young and child population.
- IT crimes are highly damaging to the environment.

Information Technology and Human Health

- Eye witness proves that spoils the eyes by radiation.
- Makes the human being corrupt and unethical practices.
- Tons of non-biodegradable plastics and components injurious to human health.
- IT Support mental injuries too by propagation of various data, video, pictures, postures, etc.
- Domestic e-waste including computers, refrigerators, televisions and mobiles contain more than 1000 different toxic material.
- Chemicals such as beryllium found in computer motherboards and cadmium in chip resistors and semiconductors are poisonous and can lead to cancer.
- Chromium in floppy disks lead in batteries and computer monitors and mercury in alkaline batteries and fluorescent lamps also pose severe health risks.
- Officials of the state run pollution board have woken up to the debilitating risk "Most of the industry, especially the Information Technology (IT) companies are vaguely aware of the problems of e-waste".

Case Study

eWaste in Bangalore

E-waste refers to discarded electronics including computers, printers, mobile phones and other such hardware.

A time bomb is tickling in Bangalore, India's hi-tech capital, but most of its six million inhabitants are largely unaware of the threat.

Home to more than 1200 foreign and domestic technology firms, Bangalore figures prominently in the danger list of cities faced with e-waste hazard.

"The end effects of electronic junk generated from obsolete computers and discarded electronic components are disastrous to our environment and people", says Mr. Sreenath, acclaimed as an e-waste expert.

"If we don't act now, we will have a polluted environment and lots of disabled children in the future," he warns. eWaste is like a slow poison.

Alarmed by the electronic pile up, Mr. Sreenath has set up Asia's first e-waste crematorium, Indian Computer Crematorium, India's technology hub.

"Here we neither bury nor burn waste. It is done through a mechanical dry recycling process'" he says.

Information Technology Participation

The Pollution Control Board has initaited talks with top companies based in Bangalore, including Infosys, Wipro and IBM.

As many as 1000 tons of plastics, 300 tones of lead, 0.23 tons of mercury, 43 tons of nickel and 350 tons of copper are annually generated in Bangalore.

More than 300 small industrial units operate in metal extraction of waste from dumped computers.

Hundreds of recyclers concentrates on discarded computers and electronic components across India and also sell second-hand parts to private computer assemblers.

Most of the recyclers work with bare hands and extract precious metals and also crude chemical processes.

Questions

1. Critically analyse the growth of population since the inception of human on the earth?
2. What are the causes for population explosion? Comment on Indian contribution for the same.
3. What is a family welfare program? How does it is done in India and its effect on welfare?

4. Environment is a critical component in human health-explain.

5. What is Human Right? How does it differ form other rights? What is the requirement for human right?

6. Debate on the topic "Human Right is a Fundamental Right Need to be Protected"

7. What is a value? What are the educational value and its contribution to environmental science?

8. What is AIDS? What are the causes and consequences of AIDS? How will you prevent from AIDS-er?

9. Women and Child need to be protected-Explain with relevant legal and international efforts to achieve this concept.

10. Information technology and environment play a role in human health care-explain.

The koala bear